350弄1-62号
255弄1-28号
出口
5273 3999

新

CapitaLand
CapitaLand

escape plan

MADE IN LINON
1.2²

三角地艺术园
人力车夫互助会旧址
星梦剧场
原嘉兴影剧院
里弄中的"秋叶原"
音乐谷商业复兴

连接：共享未来的公共空间实践

thisCONNECTION: SHARING A FUTURE PUBLIC SPACE SITE PROJECT

SUSAS
上海城市空间艺术季

2017上海城市空间艺术季案例展
SHANGHAI URBAN SPACE ART SEASON SITE PROJECT

上海城市空间艺术季展览画册编委会 编
EDITED BY: SHANGHAI URBAN SPACE ART SEASON ALBUM EDITORIAL COMMITTEE

上海·同济大学出版社 TONGJI UNIVERSITY PRESS

学术委员会

为确保空间艺术季的专业性、学术性和国际性，特成立空间艺术季学术委员会，为空间艺术季各项核心工作提供专业指导和评审意见。委员会体现国际性和专业性，委员均为国际和国内专业造诣高、工作经验丰富且有影响力的专家，并保证国际专家占一定的比例。委员会体现了多专业的跨界合作，充分体现空间艺术季是专业性、公众性相结合的城市大型公共活动。

职责

学术委员会职责包括召开各阶段学委会会议，审议有关工作事项，进行策展人和策展方案评审等；活动期间参与论坛和研讨会，进行主旨演讲和学术交流等；出席空间艺术季期间各重要节点活动等。

委员

学术委员会委员由包括城市规划、建筑、景观、公共艺术、传播、评论、策展、社会学、媒体等领域的专家组成，2017上海城市空间艺术季学术委员会委员共 31 人。

ACADEMIC COMMITTEE

To ensure SUSAS as a professional, academic and international event, the academic committee is established to offer professional guides and evaluative opinions on various core tasks of the art season. Committee members, domestic and abroad, feature expertise, experiences and influence, and a considerable number of them are internationally renowned. The committee is comprised of members from multiple disciplines, fully demonstrating SUSAS as a major urban public event that is both professional and open.

RESPONSIBILITY

Responsibilities of the academic committee include: organizing academic committee sessions at each stage where progress, curators and plans are discussed and evaluated; attending forums and seminars during the art season, giving keynote speeches and participating in academic dialogues; attending various key activities during the art season, et cetera.

MEMBERS

Members of the academic committee cover 31 professionals in urban planning, architecture, landscaping, public art, communication studies, art criticism, curation, sociology and the press. The list is subject to change according to each year's requirements.

学术委员会主任	PRESIDENTS
郑时龄	ZHENG Shiling
吴为山	WU Weishan

学术委员会委员	COMMISSIONER
按姓氏笔画排序	*SORTED BY SURNAME STROKES*
于海	YU Hai
王林	WANG Lin
王才强（新）	Heng Chye Kiang (Singapore)
王建国	WANG Jianguo
北川富朗（日）	Fram Kitagawa (Japan)
朱子瑜	ZHU Ziyu
伍江	WU Jiang
孙玮	SUN Wei
芭芭拉 · 菲舍尔（加）	Barbara Fischer (Canada)
李磊	LI Lei
李向阳	LI Xiangyang
李振宇	LI Zhenyu
杨劲松	YANG Jinsong
杨奇瑞	YANG Qirui
汪大伟	WANG Dawei
沈迪	SHEN Di
张杰	ZHANG Jie
张永和	Yung Ho Chang
张宇星	ZHANG Yuxing
郑佳矢	ZHENG Jiashi
郑培光	ZHENG Peiguang
赵宝静	ZHAO Baojing
俞孔坚	YU Kongjian
莫森 · 穆斯塔法维（美）	Mohsen Mostafavi (USA)
殷小烽	YIN Xiaofeng
曹嘉明	CAO Jiaming
鲁晓波	LU Xiaobo
曾成钢	ZENG Chenggang
戴维 · 汤普森（美）	David Thompson (USA)

前言 连接·共享未来的公共空间实践

PREFACE THISCONNECTION: SHARING A FUTURE PUBLIC SPACE SITE PROJECT

上海城市公共空间设计促进中心
SHANGHAI DESIGN & PROMOTION CENTER FOR URBAN PUBLIC SPACE

上海城市空间艺术季（SUSAS）是一项将展览与实践相结合的大型城市活动。在上海深入推进城市更新、引导城市高品质发展的关键时期，空间艺术季践行世博精神，通过推动和展示近年来城市空间的实践行动和广大市民的热情参与、积极互动，实现“城市让生活更美好”。

空间艺术季由主展览、实践案例展、联合展和公众活动（SUSAS学院）构成。实践案例展作为空间艺术季的重要组成部分，是在空间艺术季主展之外，由全市各区积极响应与参与的多元演绎与呈现。

实践案例展充分体现空间艺术季聚焦各个城市公共空间，生动呈现公共空间本身的设计水平和艺术魅力，努力推动更多城市空间更新，提升空间文化艺术内涵的活动特色。

实践案例展遴选更多新近完成或正在更新的特色空间项目，通过具有在地性的展览展示、互动体验等，生动地呈现公共空间变迁的历程，引导周边市民感受空间的变化，体会生活品质的提升，并为更多的城市规划和城市建设提供可推广和借鉴的优秀案例。

2017实践案例展以社区更新和治理、历史风貌保护和再利用、黄浦江两岸贯通等城市热点为方向，最终策划并完成了3大板块共8个实践案例展。生活与艺术，从细微处提升城市品质的社区空间——“社区，让生活更美好：浦东新区缤纷社区实践案例展”、“新泾·新境：长宁区北新泾街道实践案例展”。历史与当下，重新焕发生命活力的历史风貌空间——“‘符号上海’·上海风貌保护道路文化之旅：上海城市规划展示馆实践案例展”、“为风貌而设计：徐汇区岳阳路实践案例展”、“双重曝光·旧厂新生：静安区新业坊实践案例展”。断裂与连接，滨水等城市公共空间的复兴——“‘双向辐射’·城市更新在西岸：徐汇区西岸实践案例展”、“融‘汇’贯通：徐汇区T20大厦实践案例展”、“‘城市微介入·音乐奏响社区’：虹口区音乐谷实践案例展”。

组织、推进工作包括向各区规土局开展动员和申报，由各区规土局牵头组织策展方案编制，策展方案提交学术委员会审议，实践案例展名单确定并公布，实践案例展实施完成，顺利开幕等工作环节。

以上8个实践案例展，从策展角度来看，有些是以展览为主要形式，而有些则以空间体验为主要形式，还有些则是将展览和活动相结合。如“融‘汇’贯通：徐汇区T20大厦实践案例展”从城市建设发展、艺术文化发展的实际出发，主要通过作品展览，展出徐家汇城市空间若干实践案例和不同领域艺术家的作品；“双重曝光·旧厂新生：静安区新业坊实践案例展”通过“咱们工人有力量”和“咱们人人有力量”两期艺术活动，配合艺术装置、艺术市集和音乐派对等具有高度参与性、互动性的形式，使观展的市民能切身感受到冶金机械厂的转型之路；“‘符号上海’·上海风貌保护道路文化之旅：上海城市规划展示馆实践案例展”由上海城市规划展示馆策划的系列展览、户外集装箱展示活动、市民竞赛活动和主题系列讲座四部分组成。

正如前文提到的，对于实践案例展我们更多关注的是实践案例展的展场或是项目本身。因此，本书将通过对8个项目的历史追溯、设计理念、建设实践，以及通过实践案例展这一活动市民的体验反馈等方面，记录这些被忽视的公共空间的时代的印记，留存温暖而又鲜活的生活之本。

Shanghai Urban Space Art Season (SUSAS) is an extensive urban event that combines exhibitions and site projects. In the key period when Shanghai is deepening its urban renewal and guiding the city towards quality-oriented development, SUSAS is an implementation of the EXPO spirits, "Better City, Better Life", by promoting and showcasing recent endeavors about urban spaces and engaging citizens to participate in and interact with the program.

SUSAS is comprised of a main exhibition, several site projects, joint exhibitions and open activities (SUSAS College). As an organic component of SUSAS, besides the main exhibition, site projects also showcase various responses from other parts of Shanghai, engage and interpret the call of SUSAS in various manners and present their diversified results.

Site projects fully embody the features of SUSAS. They focus on various urban public spaces, showcase the design and aesthetical appeals of public spaces lively, encourage more urban renewal projects, promote the cultural and artful implications of public spaces.

Site projects are selected from recently completed or ongoing programs of particularities, which are then put on show to present how the involved public spaces have evolved by localized exhibitions and interactive experiences, leading the neighborhood feel the changes in spaces and improvements of life. At the same time, site projects are exceptional practices of referential and promotional values in city planning and urban construction.

The site projects of SUSAS 2017 focused on community renewal and management, protection and reutilization of historical sites, waterfront connection of the Huangpu River and other hot topics in Shanghai. Eventually, SUSAS 2017 offered eight site projects: Better Community, Better Life – Pudong Colorful Community; Xinjing, New Scape – Urban Regeneration in Beixinjing, Changning; 2017 Iconic Shanghai Serial Exhibition; Design for the Historic and Cultural Scene: Site Project of Yueyang Road, Xuhui; Double Exposure, Factory Rebirth: Jing'an Innovation Galaxy; Dual Radiation – Urban Renewal at West Bund, Xuhui; Linking Zikawei, Xuhui; and Urban Micro Intervention: Musicify the Community –Shanghai Music Valley, Hongkou.

Responsibilities of the SUSAS team include mobilizing and sending applications to urban planning and land authorities of involved districts (bureaus), helping bureaus develop curational programs, delivering curational programs to the academic committee for review, determining and publishing the list of site projects, implementing site projects and setting them up for the opening ceremony.

From the curatorial point of view, some of the eight site projects are exhibitions, some are interactive experiences while others are hybrids of both approaches. For example, Linking Zikawei focusing on the practical sides of urban construction and cultural development, is an exhibition that shows a series of cases in Xuhui District and artworks from various disciplines. In contrast, Double Exposure, Factory Rebirth is highly interactive and participatory, intended to let visitors directly feel how the metallurgical plant has evolved with its two sessions –"We Workers Have Power" and "We All Have Power" consisting of installations, art fairs and music parties. As a hybrid project, 2017 Iconic Shanghai, undertaken by Shanghai Urban Planning Exhibition Center, prepares four kinds of events: serial exhibitions, outdoor container shows, photography and painting competitions, and themed lectures.

As mentioned, we paid more attention to the sites or projects themselves when we came to site projects. Therefore this book will introduce the history, design concepts, and realisation of the eight projects and the experience carried out by the general public in order to record the often neglected public spaces of our times, to preserve the living memories for the future.

目录

CONTENTS

序一
PREFACE I

徐毅松 XU Yisong
上海市规划和国土资源管理局局长 Director of Shanghai Municipal Bureau of Planning and Land Resources

2017上海城市空间艺术季在美丽的黄浦江畔的民生码头顺利开幕。本届艺术季由上海市城市雕塑委员会主办，上海市规划和国土资源管理局、上海市文化广播影视管理局及当届主展览所在地浦东新区人民政府共同承办。

民生码头8万吨筒仓曾是亚洲最大的粮仓，随着工业时代的繁华褪去，码头已逐渐淡出人们的视野。2017上海城市空间艺术季于2017年10月15日至2018年1月15日在这里对公众免费开放，展期三个月，这个沉寂多年的工业遗址因观众的到来而重焕生机与活力。

本届空间艺术季的筹办过程凝聚了全球各界精英的智慧和心血。远道而来的策展人、参展艺术家、建筑师、规划师、各行各业的学者专家，以及一直关注空间艺术季的各方媒体，纷纷为空间艺术季出谋划策，为上海城市发展贡献自己的力量，使空间艺术季真正成为众创共享的市民文化盛宴。

芬兰建筑师伊里尔·沙里宁（Eliel Saarinen）曾说："让我看看你的城市，我就能说出这个城市的居民在文化上追求什么。"为实现2035年建设"卓越的全球城市"这一目标愿景，上海不仅要打造具有全球影响力的城市功能，还要塑造平等、包容、更富人文关怀、更具自身鲜明特质的城市文化。上海城市空间艺术季作为一个公众性的平台和窗口，将向世界展示上海独具特色的人文魅力和精彩纷呈的城市空间活力。

空间艺术季是一项将展览与实践相结合、"双年展制"的城市空间艺术展示活动。它以"文化兴市，艺术建城"为理念，以"城市公共空间品质提升"为长期固定主题，致力于通过打造具有"国际性、公众性、实践性"的城市空间艺术品牌活动，从而改善城市空间品质、提升城市魅力，实现"城市艺术化，艺术生活化"的目标。

2017上海空间艺术季由主展览、实践案例展、联合展和SUSAS学院共同构成。主展览以"连接thisCONNECTION：共享未来的公共空间"为主题，含四大主题展、12个特展、200多个展项作品，约200位来自全球17个国家和地区的规划师、建筑师、策展人、艺术家等共同参与。与上届相比，本届空间艺术季主展览希望在三个方面有所突破。一是聚焦城市空间服务于民的理念，聚焦于城市空间的连接，如大众瞩目的黄浦江滨江贯通工程、城市微更新等。二是注重以城市空间作展场，坚持展览与实践相结合，实现展示活动每举办一届，城市公共空间就美化一片，文化热点就传播一次，国内外大师作品就沉淀一批的目的。例如将主展场设置在民生码头8万吨筒仓及周边开放空间，让筒仓改建本身成为展览中最大一件展品，成为滨江新的艺术核心。三是积极搭建公众参与平台，促进公共教育，注重社会效益。既坚持学术性、思想性和前瞻性，通过国际征集，确定米兰理工大学教授、著名建筑师斯坦法诺·博埃里、同济大学建筑与城规学院副院长李翔宁、独立策展人方振宁作为主策展团队；同时又注重在市民中增加影响力，在不到两个月的时间内有超过10万的参观者，也体现了公众对于展览的关注，对于规划设计如何改变生活方式、提升生活品质的关注。

除了主展览外，全市范围内还有8个实践案例展和6个联合展，其中实践案例展包括历史风貌的保护和重塑、滨水等城市公共空间复兴、社区空间营造和更新三大类型，都是市民家门口触手可及的空间艺术和最深切感受城市温度的触手。此外，展期内还举行了100多场SUSAS学院和其他公众活动。

亚里士多德曾说："人们为了活着，聚集于城市；人们为了活得更好，居留于城市。"城市是文化的载体，文化是城市的灵魂。相比城市物质建设的速度，富有魅力的人文环境塑造也许需要经历更加漫长的过程。希望通过本届城市空间艺术季的举办，继承发扬"城市，让生活更美好"的世博精神，推动上海建设成为"更富魅力的幸福人文之城"！

Shanghai Urban Space Art Season 2017 (SUSAS 2017) is opened at Minsheng Port on the banks of the Huangpu River. It is hosted by Shanghai Urban Sculptures Committee, and organized by Shanghai Municipal Bureau of Planning and Land Resources, Shanghai Municipal Bureau of Culture, Radio, Film & TV, and People's Government of Pudong New District, where the events will take place.

The 80,000-ton Silo at Minsheng Port used to be the largest used for bulk grain in Asia, however it has gradually lost its glamor with Shanghai's industrial past. SUSAS 2017 lasts three months between October 15, 2017 and January 15, 2018, and is free to all, reinvigorating this long-forgotten industrial heritage.

This event crystalizes dite efforts and insights from countries and sectors, including curators, artists, architects, planners, researchers of various disciplines and the press. Thanks to their plans and contributions to the city, SUSAS has become a public cultural event created and enjoyed by all.

Eliel Saarinen, a Finnish architect, once said, "Take me to your city, and I can tell the cultural desires of those who live there." To become an "outstanding global city" by 2035, Shanghai needs as much an egalitarian, tolerant, humane and salient culture unique to the city as metropolitan functions of global influence. As a public platform, SUSAS aims to present the city's cultural features and vibrant urban spaces with flair.

SUSAS is a biennale that combines exhibitions and site projects, to showcase art of the space within the city. Its core idea is "Culture Enriches City, Art Enlightens Space", with its long-term theme "Improve the Quality of Urban Public Spaces". Dedicated to creating a permanent brand of urban space, it seeks to become "international, public and practical" for the sake of better urban spaces and a more charming city, with the aim of becoming a "city of art" and "art of life".

SUSAS 2017 comprises the Main Exhibition, Site Projects, Joint Exhibitions and SUSAS College. Guided by the theme "thisCONNECTION: Sharing a Future Public Space", the Main Exhibition includes four theme-related sub-exhibitions, twelve special pavilions and more than two hundred items, involving about 200 planners, architects, curators and artists from 17 countries and regions. It aims to focus on three aspects under-represented in the last SUSAS. The first are innovative ideas that urban spaces may serve the citizenry, particularly in how to connect various spaces. Prominent examples are the Waterfront Connection Project of the Huangpu River and urban micro-regenerations. The second is turning the existing urban space into exhibition sites, in accordance with the principle that exhibitions and practical life should be integrated. Each SUSAS has the fundamental task to beautify a series of public spaces, create a cultural hotspot and commission permanent artwork by domestic and foreign artists. For example, the 80,000-ton Silo, the largest in this exhibition, and its surrounding public spaces will be the new art nexus of the Huangpu River waterfront. The third is to build a public participatory platform as a faculty of general education and social well-being, while sticking to its natural inclination as an academic, insightful and forward-looking event. The main curators, the result of international competition, includes Stefano Boeri (professor of Politecnico di Milano and architect), LI Xiangning (Vice Dean of College of Architecture and Urban Planning, Tongji University) and Fang Zhenning (independent curator). The fact that it received more than 100,000 visitors in less than two months has shown the citizens' interest in the exhibition itself, and more generally, their concern about how urban design may alter and improve their life.

Aside from the main exhibition, eight site projects and six joint exhibitions are held throughout the city. These site projects are categorized into "Protection and Reshaping of Historical Heritages", "Reinvigoration of Waterfront Public Spaces and Others", and "Creation and Re-creation of Community Spaces", all of which are situated in residential areas so that they provide the public with an up-close experience of these works. Various public events and those of SUSAS College are held during this three-months period, with more than one hundred events offered.

As Aristotle once said, "People gather in cities first and foremost for security, then for economic opportunity, and then stay for the good life." Cities represent culture, and culture functions as spirit of the city. Compared to physical constructions, a charming, cultural environment may need more time to emerge. I hope that SUSAS 2017 will inherit and enhance the spirit of the Shanghai Expo, "Better City, Better Life", and contribute to Shanghai as a "city of cultural attractiveness, happiness and humanity".

序二
PREFACE II

吴孝明 上海市文化广播影视管理局 艺术总监
WU Xiaoming Art Director of Shanghai Municipal Bureau of Culture, Radio, Film & TV

上海城市空间艺术季以“文化兴市，艺术建城”为理念，打造具有“国际性、公众性、实践性”的城市空间艺术品牌活动，从而实现“城市艺术化，艺术生活化”的目标。正如市委书记韩正同志所强调的，要把文化看作上海这座城市的人民福祉、活力魅力所在、竞争力的本质体现。发展上海的文化应该多种举措并举，营造更浓厚的文化氛围。

近年来，为进一步提升城市空间文化品质，激活城市公共文化空间，上海市文广局围绕上海建设“四个中心”、国际文化大都市的目标，大力实施文化与城市规划、交通、商业等融合发展的战略，积极营造与上海城市更新、城市发展相匹配的城市文化。2016年发布第二个《营造上海城市文化氛围三年行动计划》以来，市文广局利用城市主要广场、绿地等公共空间，着力打造了城市广场音乐会；利用地铁站厅、通道、车厢等公共空间，引入雕塑、绘画、摄影、书法展览和街头艺人表演等公共艺术形式，打造精品地铁文化；利用南京路、徐家汇、虹桥、五角场、陆家嘴等商业集聚区域，引入美术、非遗、演艺等各类展览、展示、展演，打造文商互动的公共空间艺术，初步形成了社会主体参与广泛、文化活动丰富多彩、艺术空间随处可及、广大市民分享便利的上海公共文化氛围。

本届艺术季主展览以“连接thisCONNECTION：共享未来的公共空间”为主题，包含四大主题展、12个特展，有200多位设计师、艺术家、建筑师等国内外参展人共同参与。除了主展览外，全市范围内还有8个实践案例展和6个联合展。8个实践案例展选取与百姓生活密切相关的公共空间，从建设发展实际出发，通过策展与市民参与文化活动等多种方式，展示城市空间公共艺术改造方案及实施效果，吸引更多的市民关注城市空间的发展和变化，共同提升空间活力和品质。

在空间艺术季活动期间，市文广局继续落实“营造上海城市文化氛围三年行动计划”，开启“艺术+商业”联动发展的创新模式。引进高雅艺术项目入驻全市商圈、购物中心、特色商业街区等业态，推动商业设施由传统模式向体验经济转型，同时打造城市公共空间的文化厚度及人文情怀，提升市民文化素养。用多元社会主体的创意，去参与创造城市空间、激发全社会共同塑造、共同美化城市空间，为党的十九大胜利召开营造了良好的氛围。

我们希望广大市民用眼睛、文字、镜头，去发现、传播上海城市空间的艺术美、城市居民的生活美，共同参与到上海城市空间艺术的成果分享、问题思考和未来谋划中来。

Guided by the idea "Culture Enriches a City, Art Enlightens Space", Shanghai Urban Space Art Season (SUSAS) is dedicated to creating a permanent brand of urban space art that is "international, public and practical" with the aim of becoming a"city of art" and "art of life". As stressed by Han Zheng, the Municipal Committee Secretary of the Shanghai CPC, culture should be considered essentially a representation of the city's vigor, charm and competitiveness, and a variety of measures should be implemented to develop Shanghai's cultural enterprises and create a robust cultural atmosphere.

In recent years, in order to improve the culture of urban spaces and activate urban public cultural spaces, the Shanghai Municipal Bureau of Culture, Radio, Film & TV (SMBCRFT) has implemented a series of measures aiming to build "four centers" in Shanghai and transform the city into an "international cultural metropolitan", including a strategy to incorporate culture with city planning, transportation and business, and cultural projects in accordance with Shanghai's regeneration and development. Since the second "Three-Year Action Plan to Build Cultural Atmosphere in Shanghai" was published in 2016, SMBCRFT has preliminarily created a public cultural atmosphere featuring comprehensive participation, multitudinous events, pervasive art spaces and accessibility through a series of projects, including plaza concerts in squares and parks; metro culture, which includes sculptures, paintings, photography, calligraphy and street shows in stations, pedestrian walkways and carriages; public art spaces in core business areas like Nanjing Road, Xujiahui, Hongqiao, Wujiaochang and Lujiazui, where a variety of exhibitions, shows and performances are held, showcasing fine art, intangible cultural heritage, concerts and popular performance.

Guided by the theme "thisCONNECTION: Sharing a Future Public Space", the Main Exhibition of SUSAS 2017 comprises four themed sub-exhibitions and twelve special pavilions, which have crystalized the efforts of more than 200 designers, artists and architects from around the world and China. Apart from the Main Exhibition, eight projects and six joint exhibitions took place throughout the city. The site projects are all mundane public spaces. Based on the actual construction and thoughtful curation as well as public involvement, they're intended to demonstrate plans and effects of public art renovation, attracting more citizens to the development and transformation of urban spaces and participation in the improvement.

During the Art Season, SMBCRFT will continue to implement the "Three-Year Action Plan to Build Up Cultural Atmosphere in Shanghai", and launch an innovative pattern to interlink art and business. Many fine art projects will be introduced into business districts, shopping malls and street markets around the city to transform them from conventional retail areas to multi-dimensional interactive space, as well as a measure to promote cultural and humane values of public spaces and enhance citizens' cultural understandings. With the upcoming 19th CPC National Congress, these measures are calling forth ideas and efforts from various parties of our society for creating, shaping and beautifying urban spaces.

We hope that the public can spot and help spread the aesthetics of Shanghai's urban spaces as well as the fascination of its citizenry through their eyes, words and lens. In doing this, they share, reflect and broaden the fruitful outcomes from the event.

连接城市空间 以心点燃人们的心

A CONNECTION OF URBAN SPACE, A TORCH TO LIGHTEN

郑时龄 学术委员会主任

ZHENG Shiling Academic Committee President

城市是人类的化身，城市和城市空间是按照人们心目中的期望所塑造的世界。人们在构造城市和城市空间的过程中塑造了城市的未来，也重塑了自身。城市空间关乎城市的价值取向，关乎城市之人，关乎城市的未来。上海自2015年起举办城市空间艺术季的初衷，是打造一个以展示上海城市空间为平台的品牌活动，宣传和推广城市发展的理念，塑造城市空间，展示城市空间的优秀作品，从而推动城市的有机更新，完善城市空间品质，在城市空间的塑造过程中提倡高品质的设计，创造宜居的生态环境和愉悦的生活场所，让艺术介入空间。让空间品质和艺术感化人们，使人们成为更加高尚、文明、卓越的全球城市之人。城市空间艺术季不仅仅是展览和理论的探讨，也是实验性案例的展示和城市更新的实践，通过城市空间艺术季的建设和展示活动，上海的城市空间越加美好，城市环境越加宜居。

20世纪90年代以来，上海进入了重要的历史时期，城市空间结构和城市产业结构进行了全面的重组。上海在城市转型发展的过程中，面临着前工业时期和工业化时期遗留的问题以及新的后工业社会的问题。城区内仍然存在各个地区之间的差异，在环境、交通、文化教育、公共服务设施和医疗卫生资源方面存在不平衡。上海新一轮的城市总体规划提出了建设卓越的全球城市的目标，建设更具活力的创新之城、更富魅力的幸福人文之城、更可持续发展的韧性生态之城，塑造具有文化地域性、国际都会感的城市总体风貌，历史文化空间与当代功能的空间层积，弘扬地域文化的历史性与全球城市的当代性。

由伦敦市长办公室发起的关于城市文化和未来发展的全球性倡议《世界城市文化报告2012》指出：“世界城市在文化方面的重要性，与它们在金融和贸易方面的重要性等同。”上海作为卓越的全球城市，应当是一座国际化的大都市，一座多元的城市，一座拥有无与伦比的艺术和文化活动的城市，一座拥有国际学术和研发中心的城市，一座充满生命力的城市，一座以人民为中心的城市，一座安全和绿色的城市。作为全球城市，也应当是世界级的文化中心，具有持续创造科技成果和知识信息交流功能的文化、教育、科研中心。以此为目标，城市空间艺术季也承担了重要的文化使命。以城市之心去发现心，以城市之心点燃人们的心灵。

以“连接 thisCONNECTION：共享未来的公共空间”为主题的2017上海第二届城市空间艺术季，展现了城市空间与建筑和公共艺术。这个主题延续了第一届城市空间艺术季的主题“城市更新”，直面当前城市空间中的问题。我们的城市正面临着多种多样的断裂，空间的、社会的、文化的、历史的、代际等的断裂。长期以来，在城市外部和内部的社会、经济、文化等各种影响力的作用下，地区之间、城郊之间、黄浦江和苏州河两岸之间、人与自然之间都呈现了某种断裂的状态。不断往高空发展的高密度特大城市使人们在空间中的活动缺乏连续性，快速的生活节奏使人与人之间缺乏沟通，各社会阶层之间的隔阂也造成了断裂。为此，本次城市空间艺术季主题的提出是上海城市空间中各种“断点”的映照，将断裂转化为连接和延续，隐喻了城市历史文化的传承和发展。主展览所在的民生码头和8万吨粮仓正处于黄浦江滨江贯通的一个重要的节点，反映了黄浦江45公里滨江贯通的理念和实践，同时也是又一次以主展场的建设引领城市更新的杰出案例。

本次由意大利建筑师博埃里、中国建筑师以及艺术家李翔宁和方振宁联合策展的城市空间艺术季主展览包括了4个主题展和12个特展，全市范围内还有8个实践案例展和6个联合展，以及100多场各类公众活动。城市空间艺术季也越发吸引了社会和公众的广泛参与。

A city is an incarnation of people. As was expected by people, the city and the urban space are established. In construction, people have built the city's future and rebuilt themselves. The urban space of a city is as much about its future and citizens as its values. To set up a platform to present a city's urban space, Shanghai has held Shanghai Urban Space Art Season since 2015 to promote its vision for urban development, establishing an urban space and presenting outstanding works. The organic renewal of the city has been accelerated and quality of the urban space has been improved.
High-quality designs have been introduced through the urban space construction to create a livable ecological environment and pleasant living space. Art has been blended into spaces; the quality of space and the introduction of art enriches the understandings of everyday life for the general public. Shanghai Urban Space Art Season will not only provide exhibitions and welcome theoretical discussions, but also present experimental samples and offer urban regeneration practices. Through the construction and exhibition of the Urban Space Art Season, Shanghai would welcome a better urban space and more livable urban environment.

Since the 1990s, Shanghai has entered a significant historical period. The space and industrial structure of the city have both experienced full reorganization. Transformative development has to face issues left by the pre-industrialization period and industrialization period that emerged in post-industrial society. Districts of Shanghai still have differences and imbalances in aspects of the environment, transportation, cultural education, public service facilities and health care resources. The new Shanghai Master Plan has set up a goal of establishing a new global city: more dynamic and creative, with a charming humanity and more tenacious with its sustainable development and ecology, to establish an overall characteristic that combines regional cultural features and international metropolitan characters; to accumulate historical and cultural spaces with modern functional space; and to carry forward the historical significance of regional cultures and modernity of global cities.

The World Cities Culture Report 2012, initiated by the London Mayor's Office, has discussed urban culture and their future development, and pointed out: "World cities share the same significance in culture as in other aspects such as finance, trade and so on." As an outstanding global city, Shanghai should be an international metropolis and dynamic city, offering artistic and cultural activities with international academic and research centers. It will not only be a city of vitality, people-oriented, safe and green, but also a city with its own world-class culture, education and research centers, where continuous scientific and technological achievements are made and information and knowledge are shared openly. To achieve this goal, Shanghai Urban Space Art Season has undertaken its cultural mission. The soul of a city is a key to unlocking its mysteries, a torch to light its hidden areas.

Themed by "this CONNECTION: Sharing a Future Public Space", the 2nd Shanghai Urban Space Art Season has been held this year, presenting urban spaces, architectures and public art. Carrying forward "Urban Regeneration", the theme of the first Urban Space Art Season, this time it has faced up to issues in current urban space. Our urban life experiences have developed in various aspects, including space, society, culture, history and intergenerational relations. For a long time, under the influences of inner-city and suburban society, economics and culture, fragmented states have formed among regions, between the city and its outskirts, or between the Huangpu River and both sides of Suzhou Creek. People have been suffered from the fractures with nature. Activities form a lack of continuity for the megacity of a high density that keeps roaring to the sky, and people have little communication with others for its rapid pace of life. Gaps between social classes have also contributed to its fractures. This Urban Space Art Season proposed the theme as a reflection of "break points" among Shanghai's urban space, in hopes of transforming fractures into connections and continuity, by which the city's history and culture would be carried forward. Located at Minsheng Port, the 80,000-ton Silo, the Main Exhibition venue, stands as a joining point of the Huangpu Waterfront Connection, presenting the vision and practice of the 45km Huangpu Waterfront Connection, and displays itself as an outstanding example of leading urban regeneration by virtue of the exhibition venue.

With the work of joint curators, an Italian architect, Stefano Boeri, a Chinese architect, LI Xiangning and a Chinese artist FANG Zhenning, this Urban Space Art Season includes 4 theme-related exhibitions and 12 special exhibitions. In addition to the main exhibition, 8 site projects, 6 joint exhibitions and over 100 theme-related civil activities will also be held in Shanghai. Shanghai Urban Space Art Season has received growing participation from the public.

主题thisCONNECTION阐释·连接的多义性

AN INTERPRETATION OF "THISCONNECTION": POLYSEMY OF CONNECTION

thisCONNECTION/disCONNECTION，读音上非常接近，却是意义相反的一组词。我们借用这组对立意义的词汇来暗喻我们的公共空间生产所呈现出来的状态。在公共空间的生产、人与人关系的重组以及秩序建立的过程中，不恰当的设计或规划带来了更加深层的断裂。南非的城市街道，一边是富人区，一边是饥寒交迫的贫民窟。美国墨西哥边境的边境线，深深的高墙却预留了视线穿透的孔洞，每个周末在美国打工的墨西哥人民跑到这里与家人“见面”——隔着高墙的见面是手指的触摸和声音的传递。这些情景能够让我们感受到断裂与连接之间的动态转化的能量。如何将这种藕断丝连的修复力量挖掘出来并使之强化，从而来生产/修复我们公共空间中的断裂状态，正是这次展览所要探讨的主题——thisCONNECTION连接：连接不是简单的线程之间的连接，我们探讨的连接包含空间中身体的连接，空间中时间的连接，空间中物理的连接，空间中的社会连接，以及空间中的文化连接。

空间中的身体连接：如果我们的身体可以丈量空间，那么步行是最好的丈量方式。在步行中，我们的身体在观察，在记录，在体验，在与其他的身体发生碰触、故事以及记忆。随着社会的发展，步行空间在褪去，空间也因为速度而变得更加模糊与相似。这种模糊带来的断裂，是空间面貌唯一性的逝去。置于主展览展场中的装置《内省腔》，来自艺术家尹秀珍，用那些携带旧主人气息的衣物/纺织物重新编织，唤起人们在母体子宫内的记忆。她说：“人类最初的状态就是在母体的羊水中，我希望人们能回归这种状态，进行自我观照。因为在现代社会中，人很容易迷失，失去判断，变得茫然无从，不知道自己真正需要的是什么。”互动性很强的腔体内，一个个独特而鲜活的社会标本进入探索，从而也在个体之间形成全新连接的社会关系，创造记忆性的身体连接。

空间中物理的连接：空间自然的割裂一直是人类努力克服的天堑。而这种技术的超越所形成的连接有着直观而强大的力量。这种直观的“连接”也以丰富的形式出现：技术、结构与新材料在这种直接的连接形式里扮演了举足轻重的作用。而简单的连接在自身之外也带来了很多意想不到的结果：纽约Highline高线公园由跨越23个街区的高架铁路桥改造而来，丰富了城市界面，更融合了纽约这座城市不同的环境与记忆；吉首美术馆由上下两座桥叠合而成，像条街道的钢桥与包容画廊的混凝土桥直接围合出了美术馆的展厅……它们新颖地探讨了作为物理的桥的复合功能形式，也承载了文化发生的场所；当然简单的技术革新也给我们通向未来带来了更多可能性，包括主展览中的hyperloop给我们展示的技术带来的未来憧憬。

空间中的时间连接：皮埃尔·雅内的《记忆进程和时间概念》中有这样的阐释：“空间和存在性这两者在始终不断地相互定义、相互影响着，它们的集合是我们与世界的关系——体验。”

空间中的社会连接：空间是社会活动的载体，在承载社会性的同时，因为本身迅速而过度的膨胀所带来的各种出乎意料的屏障，形成了社会群体之间的交流障碍以及对立。如何创造真正的公平空间，使社会的各个层面都能共享空间，为社会各个层面提供真正的庇护，支持人类面对无边的都市，也成为我们这次展览讨论的重要课题。

Despite phonetic similarity, thisCONNECTION and disCONNECTION constitute a semantically polarized pair, which we use as a metaphor for the situation produced by our public spaces. As public spaces are manufactured, interpersonal bonds are reorganized and new orders are established, with deeper fractures caused by inappropriate design or planning. Behold the South African streets. On one side is a wealthy district, and on the other side is a ragged slum. Holes are left on the Mexico-United States barrier so that Mexicans who work in America can see their families on weekends. All these scenarios impose on us a new energy inspired between the struggle of connection and facture. That's what the theme of SUSAS 2017 is about: thisCONNECTION. It's about unearthing and reinforcing the tacit power of renovation that produces or renews our fractured public spaces. Here, we aren't connecting dots or lines, but making connections in the terms of physical, temporal, social and cultural aspects in spaces.

Bodily Connection: The best bodily way to measure spaces, if any, is walking. As we walk, our bodies are natural observers. Collisions, stories, memories are created as we interact with other bodies. Civilizational development propels walking spaces into background and corners, and with its modern speed, assimilates and blurs all spaces. Spaces used to be unique, yet no more because of blurring fractures. "Introspective Cavity" is an exhibited installment created by Xiuzhen Yin, a remix of collected clothes or fabrics with scents of their former owners that aims to arouse the viewer's memory in uterus. As the artist said, "The uterus is the starting point of our lives. The hope is that viewers can return to this state for self-reflection and self-examination. People easily get lost, losing judgment in modern society. They don't know what they really need." In this highly interactive cavity, a series of unique specimen of our society is probed. A viewer can establish otherwise non-existent social bonds with other individuals, and create a memorial connection with his or her body.
Physical Connection: Humans have been trying to overcome natural fractures, all these resulting in artificial connections that embody forces immediately felt and come in a variety of forms thanks to technology, structure and new materials. Surprising results are many beyond connection itself. For instance, Highline Park in New York is renovated from an elevated rail through 23 blocks that furnishes interfaces to the city and blends in its unique context and memory. Jishou Art Museum consists of a street-like steel bridge and a concrete bridge above it that houses galleries, with all the exhibition spaces situated in between. The museum explores the multi-functional possibility of a bridge as a physical construction, and is a place where cultural activities happen. Technological innovations may also open up the way to future opportunities, like how the hyperloop foreshadows tomorrow's hopes brought about by technology.

Temporal Connection: In the book Process of Memory and Concept of Time written by Pierre Janet , there is an explanation that "Space" and "Existence" define and influence each other in the constant. Their intersection is our relationship with the world – Experience.

Social Connection: Space houses our social life, yet as it swells disproportionately, it unintentionally cuts off communications, blocks communities and causes antagonism. How to create a truly equalized space for every group of society to share? How to truly shelter every member of society? How to support people faced with many metropolitans? These are all major questions on our issue list.

生活与艺术，从细微处提升城市品质的社区空间

LIFE AND ART: COMMUNITY SPACE AS MICRO-IMPROVEMENT OF CITIES

社区，让生活更美好：浦东新区缤纷社区实践案例展

BETTER COMMUNITY, BETTER LIFE:
SITE PROJECT OF PUDONG COLORFUL COMMUNITY

策展主题：社区，让生活更美好

EXHIBITION THEME: BETTER COMMUNITY, BETTER LIFE

项目简介

“社区，让生活更美好：浦东新区缤纷社区实践案例展”是基于浦东缤纷社区规划的一项案例展，对于探讨和市民生活密切相关的自治、共治、自下而上的城市更新方式有着重要的意义。展览呼应2017上海城市空间艺术季主题“连接：共享未来的公共空间”，从社区的角度阐述人与人、人与社区的沟通与连接。缤纷社区行动是由浦东新区人民政府主导，由上海市城市规划研究院和相关艺术院校共同参与，面向浦东新区陆家嘴、洋泾、潍坊、塘桥、花木5个街道开展的城市微更新活动。在专家和业务部门的指导下，开展了9项行动，共80多个项目。浦东缤纷社区规划在规划层面上取得了创新，不仅体现了区政府对民生的关切，更使社区周边的环境和设施得到了提升。

策展主题阐释

“社区，让生活更美好：浦东新区缤纷社区实践案例展”是缤纷社区项目的展示与推广。随着相关部门对社区规划的要求逐步提升，社区发展的短板也逐步显现，如缺少归属感和人情味，公共空间质量低下，公共服务设施不足，城市慢行系统不宜人，城市文化活动单调等。本次展览鼓励多方参与，实现人与人的“连接”、社区与社区的“连接”，希望通过一系列的艺术季社区活动，展现社区规划“精心设计、融合社区、点亮生活”的目标。

展示内容

此次展览包括1部宣传片、1个空间艺术展览、1+5“缤纷沙龙”(1个缤纷社区主论坛和5个街道分论坛)、1个低碳主题活动和1次摄影比赛，邀请5个街道与多方部门共同参与，旨在通过丰富的表现形式，呈现缤纷社区的规划成果。所有活动全部免费向社区开放，鼓励社区居民参加，让社区规划真正走进百姓生活，在各级政府的支持下为浦东缤纷社区规划创建一个重要的推广平台。

PROJECT INTRODUCTION

"Better Community, Better Life" was a site project based on the planning of Pudong Colorful Community, focusing on autonomy, co-governance and bottom-up urban regeneration approaches which are closely related to everyday life. Targeted at "micro-regeneration" of five sub-district communities in Pudong New Area (Lujiazui, Yangjing, Weifang, Tangqiao and Huamu), the Colorful Community was a program led by the People's Government of Pudong New Area, Shanghai, involving Shanghai Urban Planning and Design Research Institute and other art & design institutions. Nine events had been initiated under the guidance of professionals and authorities. The plan reflected concerns towards everyday life of local communities, in order to improve surrounding environment and realize the notion of "sharing" on the levels authorized by laws and regulations.

THEME INTERPRETATION

As the requirements for community planning increases gradually, the weakness of community development is becoming clear, for instance, the lack of community sense and humane touch, poor quality of public space, insufficient public facilities, unsatisfying non-motorized traffic system, and monotonous city cultural activities, etc.. This site project challenged these issues by engaging and connecting multiple stakeholders with the community, in the hope of representing the goal of "meticulous design, community fusion, and lightening up life".

EXHIBITION

This project included one promotion film, one space art exhibition, "Colorful Salons" (one main forum + five sub-forums), one low-carbon theme event and one photography competition. Five sub-district administrations and other authorities were involved to showcase the planning outcomes of Colorful Community. All events were open to local residents. Everyone was invited to take part and create a vital platform for the planning of Pudong Colorful Community.

陆家嘴、洋泾、花木、塘桥、潍坊5个地区共同组成了浦东新区的核心部分——陆家嘴—花木分区，东、南以内环路为界，西北至黄浦江，与黄浦外滩隔江相伴，与虹口北外滩、杨浦滨江两区隔江相望，优越的地理位置使得该区域成为规划的重点。陆家嘴—花木分区总面积28平方千米，占浦东新区城市化区域的14%。其中，陆家嘴部分1.7平方千米为金融贸易中心，是上海中央商务区的重要组成部分。花木地区是南浦大桥与杨浦大桥延伸的交汇点，以发展行政管理、商业批发和文化博览等高层次第三产业为主导，通过加强社会服务和高品质住宅配套，形成繁荣的文化商业中心。2001年，伴随金桥出口加工区的设立，花木从最早作为金桥出口加工区的居住配套区，逐渐成为国际人流由浦西到浦东迁徙的栖身之所。过去20年，陆家嘴—花木分区已发展成为浦东的成熟社区之一。然而，在这种快速生长更替中，空间与文化的失落也在发生。

针对开展城市总体规划和提高居民生活品质的双重需求，上海各区在市委创新社会治理、加强基层建设“1+6”文件精神的指导下，均开展了不同形式的社区治理和社区微更新行动。其中，浦东新区作为上海建设全球城市的核心承载区，势必要填补社区人文、文化记忆、公共空间、公共设施等方面存在的短板。根据《2016年上海市国民经济和社会发展统计公报》，浦东新区常住人口已达504.4万人，是上海市常住人口和外来人口最多的一个区。庞大的人口基础，伴随文化背景、教育程度和生活习惯的多样化，为浦东政府带来了相较于其他行政区更为严峻和迫切的挑战。此外，位于内环以内的浦东内城社区多在20世纪90年代的城市规划指导下建成，如今设施老化、住房拥挤、公共空间使用频率不高等问题突显，加速了“浦东新区缤纷社区（内城）空间更新试点行动计划”（以下简称“缤纷社区行动”）萌芽的出现。最终，缤纷社区行动在老旧小区集中、改善需求迫切的浦东内环地区选择陆家嘴、洋泾、潍坊、塘桥、花木五个街道，开展了与居民生活密切相关的九类公共要素微更新项目。这九项行动根据形态和范围大小，可大致分为三类：口袋公园、街角空间和运动场所等点状项目；林荫步道、活力街巷和慢行网络等线状项目；公共设施、艺术空间和透绿行动等其他项目。

2017上海城市空间艺术季的浦东新区缤纷社区实践案例展，可以看作是浦东缤纷社区活动与宣传的一次集中展现；将跨度长达一年，分布于浦东新区各个角落的多项行动通过展览、沙龙和市民活动，汇总为一个统一的实践案例展传达给大众。政府通过案例展传达出开放积极、渴望沟通的态度，缓解了政府与社区居民间信息不对等的矛盾，更为市民提供一种参与缤纷社区项目的有效途径，使市民获得参与公共事务和社区更新活动的满足感。公共空间品质提升与社区主人翁精神的培育实现协调配合，为社区治理创新铺垫了道路。

具体而言，此次展览包括1部宣传片、1个空间艺术展览、1+5“缤纷沙龙”（1个缤纷社区主论坛和5个街道分论坛）、1个低碳主题自行车骑行活动和1次摄影比赛。主题论坛和沙龙的筹办，既邀请了专家和设计师站在理论高度进行总结和畅想，也囊括了当地居委会自发的组织和讨论。对于非规划专业的普通市民来说，设计书和设计模型皆晦涩难懂，但置身真实的空间则可以快速直观地认识缤纷社区项目的设计和规划。由此，骑友爱好者自行设计了骑行路线，开展“低碳”出行活动，在骑行中近距离观察区内各个更新空间。另外，摄影

展用镜头记录了身边空间的变化，时空的对比也增加了活动的趣味性和受众广度。这皆为缤纷社区项目在36个社区的持续推广打下了基础。

从公有空间到共有空间

规划重心从空间向人的转变是缤纷社区的核心理念，即从注重大尺度下的空间规划到以满足人的需求为本的机制革新。在这种理念下，“市民”这一概念不只局限于抽象的数字和群体，而是一个个鲜活的有身份识别性的个体。工作中，按照年龄、国籍、职业等不同标准对人群进行分析，设身处地思考每个群体对社区空间的不同需求，小到社区中晾衣杆的数量与高度，大到社区跑道与公园的改造，从空间功能的完善来驱动整体社区生活品质的提升。

塘桥街道南泉休闲广场的更新，便是在这一理念指导下打造社区共有空间的实例。南泉广场是多个老旧社区的共同出入通道，周围还有菜市场、小学、沿街商铺等人群聚集地。改造前的广场功能混杂，设施缺乏，品质不高，菜市场门口堆满了自行车、三轮车，道路大半都被占据。为使广场更好地为周边居民服务，成为令居民乐于停留的休闲共有空间，上海市规划和国土资源管理局、上海城市公共空间设计促进中心专门搭建合作平台，招募了8家设计团队，同居委会合作提交更新方案。在方案前期，设计团队采纳了周边居民的建议，增设报刊栏遮阳顶棚和休闲座椅；在中期方案设计中，更新广场入口的雕塑，增设疗愈花廊。疗愈花廊中的植物花卉均由附近居民提供，日常维护也得到离此地不远的塘一小学植物护卫队的支持。社区内各年龄层人群，或在前期规划设计中贡献观察和智慧，或在后期运营维护中亲力亲为。居民从广场的旁观者和使用者，逐渐转变成广场的拥有者和管理者。

社区不只是物理空间上的概念，还包括生活在其中的人及其拥有的生活方式、知识储备。而社区中的公共空间，只有被公众真正使用起来之后，才有可能帮助创造习俗、文化、记忆乃至语言，也即哈贝马斯推崇的“公共场域”。如此一来，公有空间才能进一步成为理想的共有空间。这一转变过程的显著特征之一，就是居民从公共空间的“过客”到空间的“主人”的角色转变。缤纷社区系列中的摄影展和骑行活动，或可以看作是一种推动社区共有空间建设、实现开放社区的有效探索。

从结果导向到需求导向

公众的使用是社区公共空间价值的体现。这就要求工作方法需从以往的结果导向，逐步转变为需求导向。缤纷社区行动在关注项目落地效果的基础上，更强调项目实施过程中的多渠道沟通，以明确居民需求。如访谈、社会学问卷调查、微信公众意见收集、市民服务热线数据分析等，均是这种工作方法的体现。这区别于以往规划师、设计师仅跟街道、社区居委会、部分居民对接的做法，在相当程度上可以有效规避时间短、信息片段化的问题。另外，缤纷社区行动还借用第三方专业机构和渠道的研究数据进行分析。如陆家嘴街道的前期调研，参考了陆家嘴社区公益基金会在2015年发布的社区需求报告。科学性高、专业性强的专业成果可以帮助规划师更加快速地聚焦当地社区的短板与问题。大众投诉热线也是重要的信息来源。根据“12345”专线近三年的投诉电话记录，将其中有关空间设施老旧、社区脏乱差、街角环境等问题的地点绘制成“问题地图”，从定性研究向定量研究转型，将投诉量的大小作为项目落点的一个重要参考值。

缤纷社区行动的试点街道多面临老旧社区的共性问题，但街道之间不同的基础设施布局、迥异的常住居民构成，也导致不同街道和社区的居民需求以及由此导致的项目选定不尽相同。如前述塘桥街道的南泉休闲广场，其改造主要是为了满足周边居民日常休闲需求，同时可作为在校学生接送等候地点，因此将升旗台设计为拼装舞台，以促进空间使用的弹性和趣味性；为方便接送的家长和附近居民阅读，优化报刊栏，增加照明和遮阳；为增加社区交往的新节点，在慈善超市屋顶设计开放空间供居民使用。与塘桥不同，地处陆家嘴金融贸易区中心区域的陆家嘴街道，由于城区建成时间较早，二元结构十分明显：居民区、商务区并存，高档住宅区与老公房住宅区形成鲜明差异。包括在附近就业的白领阶层在内的居民，对生活配套、交通出行、人文品质提升都有着迫切的需求，福山路90号塑胶跑道项目应运而生。通过提升景观绿化、改善人行道铺装、增设街道家具与路灯等措施，跑道公园成为一个儿童可嬉闹、公众可休闲的微型休憩空间，兼顾了街区内不同类型居民的物质、精神双重需求。

多方参与的社区更新

在缤纷社区行动中，作为牵头者的上海市浦东新区规划和土地管理局通过建立自下而上的工作机制，帮助街道、社区居委明确项目目标，调动资源，进而吸引多方力量参与到社区更新的具体工作中来，社区共治自治的水平与能力也得到提升。这种搭平台、建机制的工作方式，被他们自己形象地总结为“店小二”。

首先，浦东新区人民政府创建“1+2”技术指导模式。“1”指的是一位导师：聘请杨贵庆教授等9位业内权威专家，根据其专长分别负责缤纷社区行动中的9项具体行动，为整个项目奠定了专业支撑基础；“2”指的是两位助手，其中一位助手来自市规划院或者区规划院的设计岗位，确保项目设计思维的完善；另外一位来自浦东新区规划管理中心、规划土地所的规划管理岗位，主要负责项目中各部门的协调对接，协助项目顺利推进。

其次，作为项目的主要操盘手，街道和居委会承担项目中大部分具体任务，并协调多方参与。一方面，借助妇联、团委等社会组织，在社区组织多彩的活动；另一方面，吸纳企业参与其中，如立邦公司免费提供了多条街道艺术墙绘所需的漆料。

此外，居民在这项行动中担任的角色多元且富有创造性，他们是规划设计的建议者，也是工程实施的监督者，更是项目成果的维护者。居民作为社区问题的“专家”，是最熟稔社区生活的长处与不足的人，他们对社区乃至社会的关心，是激发其参与社区更新项目积极性的源泉。居民的参与正是“多方参与、多元共治”的根本所在。

2017上海城市空间艺术季浦东新区缤纷社区行动实践案例展，向人们展示了一年时间内浦东内城试点街道社区治理的丰富内容，除包括塑造及维护社区空间的机制创新，还涉及对街道行政与技术管理层面的赋能尝试。在实践创新的基础上，将全年具体工作的经验归纳总结，制订《浦东新区关于创新社会治理，加强缤纷社区建设指导意见》《浦东新区缤纷社区建设规划土地管理实施细则》《浦东新区缤纷社区九项行动项目选址和设计指南》《浦东缤纷社区内城试点案例集》4份管理文件，以指导2018年缤纷社区行动在全区36个街镇全面推广开展的工作安排。缤纷社区行动不只是一次性的短暂运动，更逐渐发展成为一种以社区微更新为载体，通过空间治理的机制创新来提升社会治理水平的可持续长效性运动。当然，这离不开监督体系与纠错机制的完善，以应对创新过程中新的问题的出现。有理由相信，多元主体参与下的社区共建机制在浦东新区未来的推广过程中，必将做出更多的积极探索，取得斐然的成果。

Lujiazui-Huamu District, the core part of Pudong New Area, is formed by five parts: Lujiazui, Yangjing, Huamu , Tangqiao and Weifang. It is bounded by the Inner Ring Road in the east and south and Huangpu River in the west and north, facing the Bund, Hongkou District and Yangpu District - a superior location that makes it the focus of planning. The total area of Lujiazui-Huamu District is 28 square kilometers, accounting for 14% of the urbanized area of Pudong New Area. Further more, an area of 1.7 square kilometers in Lujiazui part is a financial and trade center and an important part of the Shanghai Central Business District. Huamu area, located at the intersection of the Nanpu Bridge and the Yangpu Bridge, is used to develop high-level tertiary industries such as administrative management, commercial wholesale and cultural exhibitions. It hopes to become a prosperous cultural and commercial center by supporting social services and high-quality housing. Since the establishment of Jinqiao Export Processing Zone in 2001, Huamu area has gradually transformed from a residential supporting area to the residence of the international people migrating from Puxi to Pudong. In the past 20 years, Lujiazui-Huamu District has developed into one of the mature communities in Pudong. However, rapid growth comes at the cost of spacial and cultural loss.

In light of both the Master Plan and life quality, Shanghai administrative regions have carried out different forms of activities, such as community governance and community micro renovations under the guidance of "1+6" policy in strengthening the grass-root construction of the municipal Party committee. Pudong New Area, as one of the core areas leading Shanghai to a highly developed global city, is bound to make up for the short board of the community, such as the lack of human touch, cultural memories, public facilities and the low quality of public space. According to the "Statistical Report of Shanghai on the 2016 National Economic and Social Development", the population in Pudong New Area was 5.044 million, which is the most populous district of Shanghai. The large population, with diversified cultural and educational backgrounds and living habits, presents a serious challenge for Pudong Administration. Besides, the inner part of the area was mostly built under urban planning in 1990s and now faced with a number of issues, such as aging community facilities, housing concentration, less living space per capita, and poor efficiency of public spaces. "Pudong New Area Colorful Community (inner city) space renewal pilot action plan" was born out in such a realistic condition. Within the inner ring of Pudong, five sub-districts (Lujiazui, Yangjing , Huaifang, Tantqiao, Huamu) have been designated as the experimental area for this project in 9 specific interventions. Among them, three were punctuated projects – pocket parks, corner space and sports venues; three were linear projects – boulevards, lively streets and athletic tracks; and another three were public facilities, artistic space, and green action.

There were three main considerations for Pudong in holding the 2017 SUSAS case exhibition. Firstly, trying to balance the information asymmetry between government and residents: it is possible to achieve effective two-way communication via the exhibition, and let residents establish awareness on public affairs. The government took a lot of measures to improve the quality of public space. Meanwhile, residents in and outside the community should also cultivate their master spirit to coordinate with government and promote community development so as to create a good environment. Secondly, the "three meeting system" enabled residents to participate in the whole regeneration process through prior hearings, coordination meetings and appraisal meetings afterwards, which helped local residents better understand the abstract documents issued by government. Finally, the exhibition could help realize the multiple linkage development and participation to improve government proposals, encourage the participation of residents, and promote the richness of colorful community activities.

One promotion film, one space art exhibition, "Colorful Salons" (one main forum + five sub-forums), one low-carbon theme event and one photography competition were held during the entire project. Among them, low-carbon riding event was spontaneously organized by a group of riders who designed their own route to explore every corner inside the inner ring. Compared to imperceptible design books or models, being in the real environment was more conducive for people to generate an intuitive impression on the project. Through the salons, experts and industry authorities summarized frontier concepts, while locals also expressed their opinions. The photography exhibition was more like recording the changes of the space around people through the lens. This exhibition not only spread the influence of the colorful community project, but also built a foundation for further promotion of this model to another 36 communities.

FROM PUBLIC SPACE TO COMMON SPACE

One of the most important changes in the concept of Colorful Community is the transformation from space-oriented to human-oriented planning. Traditional planning principle tends to focus on the functions of large urban blocks and the number of facilities matching with potential population. However, in the context of urban regeneration, people's daily demands are more crucial prerequisites to improve space quality. In addition, individuals are no longer confined to abstract symbols, but to specific identities. In this project, residents are divided into different groups based on age, nationality, occupation and other standards and they are treated with various demands, from the quantity and height of clothes poles to the transformation of athletic tracks and park, so as to perfect the space function and improve life quality.

The renovation of Nanquan Square in Tangqiao Street is an example. Nanquan Square is located at a junction shared by many old communities, surrounded by a vegetable market, a primary school, and many small shops. Before the transformation, the square was mix-used with poor environmental quality and insufficient infrastructure. At the entrance of the vegetable market, bicycles and tricycles were stacked and scattered, occupying 2/3 area of the roads. It was a public space that was clearly unable to meet the needs of the neighbourhood. During this renovation project organized by Shanghai Municipal Bureau of Planning and Shanghai Urban Public Space Design Promotion Center, eight design teams were invited to work with local committees in the effort of transforming Nanquan Square into a common space for the neighborhood. The results include shading facilities for newspaper kiosks, benches for the elderly to have a rest after meals or shopping nearby. A community garden named "Healing Garden" was built, where the flowers and plants were all contributed by the residents, and will be carefully maintained by the "plant escort team" from neighboring Tangyi Primary School. Everyone is contributing in their own way and gradually turning their roles from outsiders to the owners and keepers.

A community refers to a physical space and people living in that space with their different lifestyles. It's only when the open spaces in the community are actually used by its members, helping create customs, culture, memories even languages, that these spaces become what Habermas named as the "public field". A remarkable advantage of colorful community led by the Pudong government is that they pay more attention to the development of common space and the exploration of open communities. To some extent, the transformation from public space to common space is to enable residents to adapt to the changing roles from outsiders to owners and then it would improve the efficiency of space use.

FROM RESULT ORIENTED TO DEMAND ORIENTED

The key value in the renovation of public spaces is to reflect public use, which requires a shift from outcome-oriented to demand-oriented. Compared with previous community renovation activities, Colorful Community paid more attention to the operation of the project, and collected residents' needs through various sources, including street interviews, sociological questionnaire, WeChat public opinion collection, and hotline data analysis. Planners and designers normally start working by interviewing sub-districts, community authorities and some of the residents, but such information collection method in a short time may be fragmented and incomplete. Instead Colorful Community in Pudong collected public opinion at the very early stage using both online and offline methods. In the preliminary research stage of Lujiazui Street, through referring to the community demands report released by the Lujiazui community public welfare foundation in 2015, it was more comprehensive and clearer to understand the short board and problems of local communities. In addition, the "12345" complaint hotline in the past three years had been used to identify problems regarding old facilities, dirty and messy spaces etc. and helped draw a "problem map". As a result, from the qualitative research to the quantitative research, the corner spaces with the largest number of complaints were chosen to use as a project placement.

Despite of the common issues, the streets involved in Colorful Community activities still differ from each other in regard of infrastructure layout and resident composition. This requires a diversified working method. For example, Nanquan Square in Tangqiao Sub-district was designed to meet especially the needs of parents who come to pick up their children there: The flag raising platform was designed as a stage for parents and passersby to stay and read while waiting; lighting and shading were added for newspapers columns; open spaces were designed on the roof of Charity Supermarket to increase community interaction. Lujiazui street, however, is quite different. It lies in the central area of Lujiazui Financial Zone, where most of the neighborhoods are already well developed. The dual structure, the contrast between residential and business areas are distinct. Housing conditions, travel environment and property management in the old residential areas were poor. The integration degree in that area is not very high and needs of white-collar class do not receive enough supports on daily necessities, transportation and cultural nutrition. To meet the needs of the two groups, No.90 Fushan Road built the Runway Park into a miniature playground through improvements on plastic pavements, landscaping, furniture and street lighting. The park also hosts street salons. In the future, the community foundation will carry out activities such as "fun games" and open-air films in Fushan garden to meet the needs of diversified residents.

MULTIPARTY PARTICIPATION IN COMMUNITY RENOVATION

Pudong New Area Planning and Land Authority had established a bottom-up mechanism to help the sub-district and the neighborhood committee to clarify goals and mobilize resources, to attract more participation into the specific work of community renewal. A "1+2" technical guidance model was established for each neighborhood to have support from one expert and two assistants. Professor Yang Guiqing and other eight experts were invited to guide nine activities respectively according to their professional backgrounds. The assistants — one with design background and the other from administrative position — helped improve design thinking and coordinate with all the departments that got involved.

Secondly, as the main operators and decision-makers of the project, the sub-district and community committee undertook most of the specific tasks that attracted many parties to participate. With the support of social organizations such as the Women's Federation and Youth League committees, diverse activities were organized to strengthen mutual trust and cohesion among residents. Enterprises also participated in this project through connecting streets committee, such as the Nippon state company providing free paint for wall paintings in Pudong.

Local people have played multiple roles in the project, from contributors of proposals to supervisors of project construction, and most importantly the maintainers and recipients of future project outcomes. People who live here know the best about the missing parts of public life, therefore collaboration and multi-party governance are the best ways to implement a better community in the future.

This site project has shown the rich content and achievements of community management in Pudong's pilot neighborhoods in the short period of one year. These results include innovations in shaping and maintaining community spatial mechanisms and attempts to empower the administrative and technical staff on the street level. In order to summarize the project experience, four documents have been written ("Pudong New Area's Guiding Opinions on innovative social governance, strengthening the construction of colorful communities", "The implementation rules of land management in the 'colorful community' construction plan", "The guide on site selection and design methods of nine actions in Pudong New Area colorful community" and "Case Sets of Pudong Colorful Community Inner City") to guide the expansion of 2018 Colorful Community Project in the region's 36 neighborhoods to promote a comprehensive work arrangement. The Colorful Community is not a short term campaign, but a sustainable long-term project that promotes multi-involvement through building innovative mechanism with micro-renovation as the carrier. The growth of completed mechanism is bound to the birth of various contradictions and problems, hence the establishment of supervision system and error correction mechanism is much needed. Still, we have every reason to believe that community participation system under multiple participants will make a difference in Pudong New Area in the coming years.

采访：策展人奚文沁

在浦东缤纷社区行动中，您的团队主要参与了哪些工作？

上海市城市规划设计研究院和浦东新区规划院共同承担了浦东新区缤纷社区行动中五个街道的总体规划任务。首先，团队在社区街道向公众征询意见，寻找、分析社区现有的问题与矛盾，然后思考如何将这些问题与相应的空间对应起来，需要新增或更新哪些空间，以提升居民生活品质。在规划的具体过程中，从宏观层面思考服务半径的规模和数量如何才是合理的，在此基础上，随时与街道和居民互动反馈，不断调整修改方案的细节，使方案更加完善。

在项目进行中，您的团队和导师是怎样的合作关系？

缤纷社区的导师制度是从2016年开始实施的。简单来说，我们是以设计师为主的规划团队，而九位导师对应九项行动内容。导师都是在业界有一定知名度和影响力的专业人士，他们担任团队的引领者、监督者和支持者。很多讨论导师都会亲临指点，给出建议。

项目选定主要考虑哪些因素？

2018年缤纷社区在浦东新区36 个街镇全面铺开，因为时间有限，不可能全部做整体规划设计。因此我们的策略是让街镇成为主要的项目推介者，专业的规划团队通过实地调研，与街镇层面沟通商量并最终确认或变更项目。从规划的角度来说，项目的选择更需要关注公共空间和公共利益，因为公共财政用在更多体现公共利益的项目上，能使更多居民受惠。但在社区更新项目的选择方面，小到社区内部的楼道、道路，大到街道的公共空间，皆可推进具体的城市更新。

公众参与环节如何展开，有哪些经验？

浦东新区创立的“一图三会”，即一张蓝图的展示，佐以项目事前征询会、事中评议会、事后评估会，就是为了更好地促进公众参与。面对面的访谈、微信群等社交媒体软件的使用，确保了政府与居民对社区更新项目问题的时时沟通，事事透明。除了征求居民的意见外，更要确保居民对政府工作进程的知情权，这也符合规划转变时期开门做规划的理念。应当针对需求、针对问题去规划，否则用处不大。

在参与缤纷社区之前，我们已经在公众参与方面积累了一定的经验。一是认识到居民在很多项目内容上的专业性与参与热情，二是明确了民众参与的途径和信息获取通道的重要性，三是参与人群越来越多元，不仅仅是老年人，更多的年轻人和非盈利组织也加入进来。

谈谈您在2017上海城市空间艺术季中的策展思路？

这次展览可以看作2017缤纷社区行动系列活动的一次宣传，包括宣传片和六个论坛，还有自行车骑行活动。策展思路根据本次空间艺术季的主题“连接：共享未来的公共空间”，聚焦“社区链接”，展示缤纷社区行动中优秀的案例，以及整体的逻辑思路、理念机制。为了更好地表现社区互动，摒弃了相对深奥晦涩的专业语言表达，转而采用很多生动、朴实、喜闻乐见的形式，比如照片、卡通图片等。期间我们还举办了一个摄影展，居民参与得非常积极，作品评选有三四百幅，最后好中选优，选出100幅获奖作品并集中展示。

INTERVIEW WITH CURATOR XI WENQIN

What kind of role your team played in the Colorful Community project in Pudong?

The Shanghai Urban Planning and Design Research Institute and the Pudong Planning Institute jointly undertook the overall planning task of five sub-districts in the colorful community action in Pudong New Area. First, we consulted the public, tried to find and analyze the existing problems in the community, and considered how to match these problems to specific space and which places we can supplement or upgrade to improve the quality of life. During the process of planning, we had to consider from the macro level to set a reasonable scale and number of service sizes. On the other hand, we also needed to interact with the sub-districts and residents at any time, and constantly adjust and modify the details of each scheme.

How does your team work with the mentor during the project?

Mentorship system was first introduced last year as an innovative practice. We are a planning team grouped by designers and nine mentors who were responsible for nine actions. All these mentors have their reputation and influence in their specific filed, and they gave us advice whenever we discussed about the project, like supervisors or supporters.

What are the main considerations for the selection of the project?

The colorful community project will fully spread in 36 towns and sub-districts in 2018. Due to the limited time, it is impossible to carry out the overall planning of all the projects. Therefore, the sub-districts took charge of promoting the project, and we did the field research at the town level, to confirm the project. From the perspective of urban planning, more attention should be paid to public space and public interests. Government funds should be used in solving public issues to benefit more people. However, in terms of the choice of community regeneration projects, as small as the corridors and roads within the community, and as large as the public spaces on the streets should be included.

How did you involve public participation?

This is done through "One Blueprint, Three Meetings", that is a prior consultation meeting, an in-process appraisal meeting, and an evaluation meeting. We organized face-to-face interviews and forums, also kept locals updated via social media. In addition to seeking public opinion, people's right of knowledge should also be protected. Urban planning should be open and be communicated to the general public. It's useless unless it's demand-oriented.

We have been doing public engagement for years and accumulated some experience. The first thing we learned is that residents are more professional and active in community planning than we think. Second, it's very important to provide people with access to participate and access to information. We also realized participants of community renovation are getting more diverse. No longer limited to elderly people, more young residents and non-profit organizations are getting involved.

Could you talk about your ideas for curating this exhibition?

This exhibition is actually part of our original colorful community series, which includes a promotional clip, five or six forums, and cycling activities. At 2017 SUSAS, we mainly demonstrated the best cases and the overall planning logic and ideas. According to the theme of this art season "to connect the future", the theme of our community case exhibition is "community linkage", which means link community to people, as well as connections among individuals. We used vivid, honest and popular forms of expression to connect with local people such as photos, cartoon, etc. We even held a photo exhibition. People were very active and we received three or four hundred photos in the end, 100 of which were selected for the exhibition.

展览基本信息

主题：社区，让生活更美好

时间：2017.10.15—2018.1.15

举办地址：民生码头八万吨筒仓7楼

主办单位：上海市浦东新区人民政府

承办单位：上海市浦东新区规划和土地管理局

协办单位：上海市城市规划设计研究院；上海市浦东新区规划设计研究院；中国（上海）自由贸易试验区管理委员会陆家嘴管理局；陆家嘴街道；潍坊街道；洋泾街道；塘桥街道；花木街道

策展人介绍

奚文沁

高级工程师，现任上海市城市规划设计研究院规划二所所长，历史文化和公共空间艺术中心主任。2015上海城市空间艺术季主展览上海特展板块策展人。先后参与或主持完成的规划设计或研究项目近千项，多次获得全国、上海市优秀规划设计奖，其中"黄浦江两岸公共空间贯通开放规划"获得2018年全国优秀规划设计一等奖。

徐继荣

2013年毕业于同济大学城市规划专业，现供职于上海市城市规划设计研究院规划二所，参与并主持完成各类规划设计与研究课题。2010年参与的合作项目《老有所乐》论文获全国高等学院城市规划专业学生社会调查报告一等奖。曾参与2015上海城市空间艺术季主展览上海特展板块策展。

杨帆

2012年毕业于伦敦大学学院巴特莱特建筑学院，2013年3月进入上海市城市规划设计研究院规划二所工作。

文

刘雅妮[城市中国]+崔国[城市中国]+赵波[上海市浦东新区规划和土地管理局]

采访

丁馨怡[城市中国]+刘雅妮+崔国

图片提供

徐继荣

Basic Information

Theme: Better Community, Better Life

Time: 15 October, 2017 – 15 January, 2018

Venue : 7/F, the silo of 80,000 tons MinSheng port (No.3, Minsheng Road)

Host: People's Government of Pudong New Area, Shanghai

Organizer: Pudong New Area Planning and Land Authority

Co-organizers: Shanghai Urban Planning and Design Research Institute; Pudong New Area Planning and Design Research Institute; Lujiazui Office of China (Shanghai) Pilot Free Trade Zone Administration; Sub-district offices of Lujiazui, Weifang, Yangjing, Tangqiao, Huamu

Curators

XI Wenqin

Senior engineer, director at No. 2 planning Dept. and Historical Cultural and Public Space Art Center. She has participated in or led nearly 1000 planning, design or research projects, and won a number of national or municipal awards of planning or design. In recent years, she has focused on the work around the World Expo, the protection of people's livelihood, heritage culture. Her awarded projects include "Huangpu River Waterfront Public Space Planning" (2018 National Outstanding Planning and Design Award); "Hengshan Road – Fuxing Road Historical and Cultural Heritage Protection Plan" (first prize, Shanghai Municipal Planning and Design Award). She was the curator of Shanghai Special Exhibition in SUSAS 2015.

XU Jirong

Graduated from Tongji University in 2013 with master degree in Urban Planning, and started working in Shanghai Urban Planning and Design Institute in September in the same year. She now works at No. 2 planning Dept. She has participated in and led various planning, design and research projects, with experience as planner and designers.

In 2010, she contributed to "The Old's Happiness", which won the first prize of National Outstanding Social Investigation Reports of Urban Planning Students. She helped curate the Shanghai Special Exhibition in SUSAS 2015.

YANG Fang
Graduated from UCL Bartlett School of Architecture with a master degree, and has been working in No.2 Planning Department, Shanghai Urban Planning and Design Institute since March 2013.

Text
LIU Yani [Urban China] + CUI Guo [Urban China] + ZHAO Bo[Pudong New Area Planning and Land Authority]

Interview
DING Xinyi[Urban China] + LIU Yani + CUI Guo

Image Source
XU Jirong

九项行动在浦东新区的分布示意
Distribution of nine interventions in Pudong New Area

空间艺术展主展场现场，浦东实践案例展板块整合了
5个街道的项目成果，统一展示
The exhibition presented outcomes from the five streets

街道论坛以最大程度调动社区居民积极性为首要目的
Street forums were more focused on stimulating the enthusiasm of community residents

“低碳出行”主题活动，用脚步丈量社区的变化
Low carbon outing events were organized for people to see the changes in the community

主论坛以相对专业、正式的形式，探讨缤纷社区的机理与愿景
The main forum explored the theoretical feasibility of communities in a relatively professional form

改建后的南泉休闲广场为附近居民提供了休闲阅读好去处
The renovated Nanquan Square provides a good place for residents to read and relax

跑道公园丰富了人行道的景观设计
Runway Park is an enhancement to the walkway landscape

缤纷社区行动的“缤纷”明信片
The matching postcards of the Colorful Community

浦东缤纷社区行动中，街角的彩色墙绘为传统的街道空间增加了一丝灵动
Wall paintings in the street landscape

社区文化活动中心成为居民学习、消遣的好去处
Community Cultural Activity Center enriches the daily life of local residents

新泾·新境·长宁区北新泾街道实践案例展

XINJING, NEW SCAPE: SITE PROJECT OF CHANGNING BEIXINJING COMMUNITY

策展主题：新泾·新境

EXHIBITION THEME: XINJING, NEW SCAPE

项目简介

北新泾街道位于长宁区中北部，由新渔港—淞虹路—苏州河—元丰天山花园—天山西路—北虹路围合而成，辖区面积1 403 500平方米，前身是具有700多年历史的北新泾镇。北新泾街道现状以老旧居住小区为主，社区环境有待提升，公共服务设施与公共活动空间活力不足。近两年，街道希望通过社区内的微更新，有针对性地提升环境品质，完善服务设施。至2017年10月，累计完成22个更新项目，类型覆盖城市系统更新（天山西路景观改造）、城市公共空间改造（平塘路金钟路口街角广场改造等）、社区服务设施更新（新泾七村三亭改造等）及公共艺术项目（金钟路围墙改造等）。“新泾•新境：长宁区北新泾街道实践案例展”是对这些微更新试点的全面梳理与展示。

策展主题阐释

“新泾•新境”案例展从三个层面诠释2017上海城市空间艺术季“连接：共享未来的公共空间”的主题内涵：一是以道路、桥梁为代表的城市物理空间的弥合；二是社区空间在生活方式和使用方式上的重置与激活；三是居民与城市议题之间的对话。展览试图在基础建设和生活方式这两个层面上回应一种“连接”，同时展览本身也应当成为一种沟通的渠道，使城市空间的概念真正走入街道，进入社区。

展览旨在展示长宁区北新泾街道2016—2017在城市更新层面所做的实践，以及这些实践为社会带来的全新环境与认识，这也是展览以“新境”命名的缘由。在经历了为期两年的社区空间微更新之后，将微更新话题从小型公共设施的改造，延伸到城市系统的营建——街道界面、道路断面、公共设施这些与居民生活息息相关的内容。北新泾街道因其城市空间上的典型性，将为人们开启另一种全新的观察视角。

展示内容

长宁区北新泾街道实践案例展主要由“更新进入新泾”“展览进入社区”“学术进入日常”三大板块组成。其中“更新进入新泾”以案例为主，集中展示微更新与城市系统更新两方面的举措与设想，前者集中展示一批2016—2017年的试点工程，包括金钟路平塘路改造、新泾五村社区中心改建等；后者则是针对天山西路改造与北新泾街道整体提升的建议。“展览进入社区”代表本次的展览空间融入社区，一方面用步行流线串联起历年试点，另一方面将开闭幕式的展场设置在社区中。“学术进入日常”在展览期间举行了多个专题活动，邀请北新泾居民与城市、建筑、规划、社区营造方面的专家共话发展。值得一提的是，展览期间，作为案例展主展场的长宁民俗文化中心正经历改造更新，处于施工状态下的场馆作为更新案例中的一则，是整个展览乃至区域的缩影。

PROJECT INTRODUCTION

Located in the north-central part of Changning District, Beixinjing Sub-district is enclosed by Xinyugang, Yuhong Road, Suzhou River, Yuanfeng Tianshan Garden, Tianshan West Road and Beihong Road, covering an area of 140.35 hectares. It used to be Beixinjing Town with a history of more than 700 years. The current Beixinjing Sub-district is dominated by old residential quarters, where the community environment needs to be improved, and public service facilities as well as the vitality of public activities spaces are insufficient. In the past two years, the sub-district hoped to improve the environmental quality and the service facilities through micro-regeneration measures in the community. By October 2017, 22 regeneration projects had been completed, covering urban system renewal (Tianshan West Road landscape reconstruction), urban public space renovation (corner square renovation at Jinzhong Road and Pingtang Road, etc.), community service facility renewal (three pavilions renewal in Xinjing Qicun, etc.) and public art projects (wall renovation at Jingzhong Road, etc.). "Xinjing, New Scape - Site Project of Urban Regeneration in Beixinjing Changning" is a comprehensive review and demonstration of these micro-regeneration pilots.

THEME INTERPRETATION

This site project interprets the theme of the 2017 Urban Space Art Season "this CONNECTION: Sharing a Future Public Space" from three aspects: First, the physical connection of urban space represented by roads and bridges; Second, the reset and activation of the community space in terms of lifestyle and usage; Third, the dialogue between residents and urban issues. The exhibition attempts to respond to "this connection" at both the infrastructure and lifestyle levels. The exhibition itself should also be a communication channel that allows the concept of urban space to truly enter the street and community.

"Xinjing, New Scape" Site Project intends to demonstrate the practice of urban regeneration in the past two years in Changning District, as well as the new environment and new understanding these practices have brought to the society. This is also the reason why the exhibition is named as the "New Scape". After a two-year micro-regeneration of community space, we believe that the micro-regeneration topic should be further extended from the transformation of small public facilities to the construction of urban systems: street interface, road sections, public facilities and others closely related to residents' daily life. Beixinjing Street will open another new observation dimension and perspective due to its typical urban space.

EXHIBITION

"Xinjing, New Scape" Site Project mainly consists of three parts: "Regeneration into Xinjing", "Exhibition into community" and "Academics into everyday". "Regeneration into Xinjing" is mainly based on cases, focusing on the initiatives and ideas of micro-regeneration and urban system renewal. The former shows a batch of pilot projects from 2016 to 2017, including renovation of Jingzhong Road and Pingtang Road, the renovation of the community center in Xinjing Wucun; the latter is the proposal for the renovation of Tianshan West Road and Overall improvement of Beixinjing Street. "Exhibition into community" represents the integration of these exhibition spaces and the community. During the exhibition, the pilots over the years are linked by the walking system on the one hand, and on the other hand, the opening and closing ceremonies are set in the community. "Academic into daily" holds various events during the exhibition, inviting local residents to develop together with experts in urban study, planning, architecture and community building. It is worth mentioning that during the exhibition opening period, the main venue Changning Folk Culture Center was undergoing renovation and renewal. The venue under construction was used as one of the regeneration cases, which became the epitome of the entire exhibition and even the region.

南宋建炎至祥兴（1127—1278）年间，因地处水路交通要冲，新泾北通吴淞江（苏州河）处出现集市，后集镇称之北新泾。明末清初，这里先后出现小金更、曹家宅、赵巷等22个村落。1911年，新泾同虹桥等其他四地合并为蒲松镇，并于次年改为蒲淞市。至1926年归淞沪商埠督办公署管辖，建蒲淞区。20世纪30年代，天原化工厂等工厂在此创办，商业日趋繁荣，成为沪西农副产品集散地，同时又为棉花集市，南北货物在此运输交易，有了“北码头”的称呼。1938年，北新泾镇与华漕镇曾合并为新华镇，后于1949年前夕归属上海市新泾区管辖。1949年5月24日北新泾镇解放，成为新泾区第一办事处。此后至1984年，北新泾镇行政归属数次变动后，最终划入长宁区，于1996年改镇为街道。北新泾街道位于长宁区西部，总面积2.22平方千米，由于区域的工业发展，20世纪80年代开始一部分居民从市区动迁至此，北新泾街道如今80%左右的房屋也是在其间建成。因此，老旧高密度居住型社区在新的发展背景下，亟须获得匹配现有需求的规划突破。社区更新是这次突破的第一步。

社区作为城市治理的基本单元，是城市更新的重要载体，反映了城市作为复杂生命体的诸多特征。“行走上海：社区空间微更新计划”专项活动自2016年开始实施，由上海市规划和国土资源管理局组织开展。这一计划的更新对象，经过前期社会征集和实地调研筛选，最终选择与居民生活密切相关的11个社区公共空间作为试点项目，涉及6个区8个街道。相较于2016年和2017年的更新计划多关注老旧小区公共空间的方向，2018年微更新走出老旧小区，探索新的空间类型，走向城市各类灰色空间，比如容易被人忽视的桥下空间等。从更新策略来讲，社区微更新主要是通过区规土局、街道、社区公益组织、设计师、艺术家和市民等多方群体参与搭建的平台，群策群力，在社区公共空间微更新方面进行实践与创新。当下的微更新不仅激发了公众参与的积极性，更通过协调联动政府、专业机构、设计师、企业、居民等多方主体，在某种意义上推动了政府自上而下的“有意”放权，反向引入“上下结合”的合作模式，是社区管理机制的一次创新，也是城市治理的一次突破。

作为2017上海城市空间艺术季实践案例展的一部分，北新泾街道实践案例展代表社区微更新的成果，以“新泾•新境”表达社区营建的理想。1984年，北新泾镇正式划入长宁区，随后这里逐渐成为上海著名的大型人口居住区。由于建造年代较早，小区内建筑密度高，公共空间狭小，基础设施建设相对落后，亟待更新。自2016年开始至2017年10月，街道执行了22项社区微更新项目措施，覆盖城市更新的多个门类和层级，有针对性地提升环境品质，完善服务设施。

为了回应主展场“连接”的策展母题，北新泾案例展试图从三个层面理解本届上海城市空间艺术季的主题：一是以道路、社区入口等为代表的城市物理空间的弥合，二是社区空间在生活方式和使用方式上的重置与激活，三是居民与社区营造、城市建设之间的对话。案例展试图从空间、使用上建立起连接，同时展览本身也应当成为沟通的渠道，让城市空间的概念真正走入街道，走入社区，走入生活。主展场选址在长宁区民俗文化中心，作为策展案例中的重要一环，文化中心在策展团队的改造下，完成了一次社区更新的“教育展”。策展内容梳理了北新泾社区实施的微更新项目，2016年北新泾街道启动公共空间微更新改造，涉及8个社区空间微更新项目，2017年增加了多个微更新试点，是多方参与下的对于社区营建的积极尝试。

案例展通过“更新进入北新泾”“展览进入社区”“学术进入日常”三大板块，探讨社区更新的使命和愿景。“更新进入北新泾”提供了社区更新的实施案例；“学术进入日常”，通过多个专题活动邀请社区居民与专家探讨社区未来；“展览进入社区”则是展示北新泾街道空间的全貌，回顾了过去两年北新泾街道在社区更新层面的尝试，将更新议题从小型公共设施改造延展到城市系统改造，为上海城市空间的提升提供全新的观察与思考。

As the basic unit of urban governance, community is an important carrier of urban renewal, reflecting many symptoms of the city as a complex living organism. Since 2016, the special event "Walking in Shanghai — Community Space Regeneration Plan" has been developed by Shanghai Urban Planning & Land Authority. The updated objects are 11 community public spaces selected from the previous societal recruitments and field surveys, whose subject matters are closely related to residents' lives. They involved 6 districts and 8 streets. Compared with the focus on open spaces in old communities between 2016 & 2017, the regeneration plan in 2018 explored new types of space including various gray areas in the city, such as space under bridges which can be easily overlooked. In terms of the strategy of regeneration, the community micro-regeneration can be accomplished through the platform made up of the local official department, the street office, non-profit organizations, various designers, artists and citizens who participate and work together to achieve the practice and innovation on micro-regeneration in community. The current micro-regeneration has not only stimulated the willingness of the public to participate, but also coordinated the power of the government, professional institutions, designers, enterprises, residents and others. To some extents, it has encouraged the government to "deliberately" decentralize from top to bottom, and introduce the "up and down" cooperation model adversely, which is an innovation of the community management mechanism and also a breakthrough in urban governance.

As a part of the "2017 Shanghai Urban Space Art Season" site project, "Beixinjing Site Project" represents the outcome of community micro-regeneration, and expresses the hope of community building through "Xinjing, New Scape". Since 1984, Beixinjing town has been officially included in Changning District, and has gradually become a well-known residential area in Shanghai because of its large population. Due to the early construction, the building density in the community is high, the public space is narrow, and the infrastructure construction is relatively backward, which calls for a new round of regeneration. Since 2016, the street has been expected to improve the environmental quality and the service facilities through the micro-regeneration measures in the community. By October 2017, 22 regeneration projects had been completed, covering urban system renewal (Tianshan West Road landscape reconstruction), urban public space renovation (corner square renovation at Jinzhong Road and Pingtang Road, etc.), community service facility renewal (three pavilions renewal in Xinjing Qicun, etc.) and public art projects (wall renovation at Jingzhong Road, etc.).

In response to the general topic of the main exhibition "this CONNECTION", Beixinjing site project attempts to understand the theme from three aspects: First, the connection of urban physical spaces represented by objects such as roads and bridges; Second, the reset and activation of lifestyle and use in community space; Third, the dialogue between local residents and urban issues. The site project tries to establish a connection between infrastructure and lifestyle, and the exhibition itself should also be a channel for communication that allows the concept of urban space truly come down to earth into the street and the community. The main exhibition site is located in Changning District Folk Culture Center. As an important part of the case study, the Cultural Center completed an "educational exhibition" about community regeneration under transformation. The site project went over the development of micro-regeneration projects in Beixinjing community. Beixinjing street launched micro-regeneration of open spaces in 2016, including 8 projects. In 2017, a few micro-regeneration pilots have been added. It is a positive attempt for community building with multi-participation.

"Xinjing, New Scape" Site Project explores the mission and prospect of community regeneration through three sections "Regeneration into Xinjing", "Exhibition into community" and "Academic into daily". "Regeneration into Xinjing" provides practical cases of community regeneration; "Academic into daily" invites community neighbors and experts to discuss the community future through various special events; "Exhibition into community" shows the overall picture of Beixinjing Street and reviews the attempts that Beixinjing Street has taken in the community regeneration in the past two years. What's more, it extends the regeneration topic from small-scale public infrastructure renewal to the transformation of urban system, providing new observation and thinking for exploring Shanghai urban space. By interviewing two co-curators and the head of Beixinjing Street regeneration project, this article attempts to provide reflections of community regeneration, as an important ecosystem in urban operations, from different roles and perspectives.

采访：
策展人张佳晶&周渐佳+张生义（北新泾街道办事处主任）

微更新，释放城市空间的有效途径

针对北新泾实践案例展，在展场选址、展陈内容等方面，面临哪些棘手的难点并如何应对？

策展人：首先从城市更新的前置条件来看，长宁区几乎没有新供应的土地，主要在做存量的更新与盘活。所以，相较于上海市其他行政区域，长宁区的建设步伐比较缓慢，如果对标浦东新区的缤纷社区，在北新泾街道展开的社区更新铺开的面积比较小。

最初我们将主展场选址在新泾五村，当时设计了一个车棚加建的方案，形成半开放的健身场所，也不扰民，希望在开幕展结束后可以引导人们回归社区，但方案被否决了。后来了解到民俗文化中心有闲置的使用空间，但这一类型的建筑空间属性限制了布展方式，所以策展团队在保留展厅原貌的条件下做了一个空壳。当展览撤离、展板拆除后，像是“什么都没有发生”。具体到展览内容，除了街道上报的22个参展案例，这次展览主要延伸了两部分内容：一是天山西路的系统更新，二是置入以城市空间为讨论命题的皮影戏。

“皮影戏”的类艺术装置主要反映了城市更新中的哪些核心要素？

策展人：这出皮影戏叫作“侬当心伐要拐高”（沪语，你要小心不要摔跤）。选择皮影戏的缘由是恰巧在展场看到演员们拿着皮影戏用的布偶，就试图用皮影戏编写关于城市空间、街道空间、城市生活的短片。剧本上隐去了一部分内容，比如拆违。我们还邀请一些职业演员，诠释城市生活，以视频的形式在展厅循环播放。我们营造了在施工状态中的展场进行展览的氛围，可以视作“更新中的‘更新展’”。它是一种“三步展”，也就是在更新的基础上做教育，通过皮影戏里的人物对话，反映城市生活的核心，而不是简单地在室内空间的表皮上“涂脂抹粉”。

天山西路系统更新项目最终没有落地，但通过这次策展，我们展出了道路系统的更新内容，这些内容可以反馈给长宁区政府。创造一种新的认知，比如道路的转弯半径可以缩小，路幅也可以缩窄等，为未来的城市更新提供更大的可能性。具体到设计上，天山西路有6条车道（包含2条人行道、2条自行车道以及隔离带），我们将3.75米宽的主车道缩窄为3米，释放出近3米的空间留给步行。其中的启示是，街道的更新不是美化路面，而是要挖掘新的潜力，比如能否将三车道转变为两车道。当然，这和城市倡导的交通方式有关，如果能够在既有的情况下将传统的车道变为更“经济”的车道，腾挪出新的空间，人行道宽度就可以增加两米，甚至还可以增加助动车道，化解机动车与自行车并行时引发的矛盾。

所以我认为城市更新的核心是释放出新的空间，而非单纯的城市美化运动。对居民而言，大到城市公共空间、城市车行空间、社区入口的功能性，小到社区公共活动场所的有无、绿化的宜人性等都是最核心的问题。

策展结果上，本次案例展是否有突破？

策展人：策展期间，我们在主展场讨论了慢行系统，闭幕时也探讨了如何做社区更新。目前来看，讨论的内容对专业领域的人群能够产生讨论价值，但普通居民可能不太关心。所以我们希望在可能的情况下做些调整，比如向大众普及专业知识，或试图挑战一些东西，给我们所生活的城市提出一些问题。另外，也希望策展的参与方能站在共同的认知层面，集中精力放大策展的可能性，而不是集中在细微的更新点位。城市系统更新，不能脱离大系统来谈微小的空间。因此策展有时需要突破一些边界，比如西岸艺术中心在策展期间被推到空间合法化的轨道上，空间能够因为一次城市空间艺术季而发生改变。

重组社区生态
系统细胞

北新泾街道社区微更新与其他城区相比，有哪些特点和制度上的突破？

张生义：北新泾街道辖区原本属于城乡结合部，80%左右的房屋都是20世纪80年代初期建成的。商品房在整个社区中的比例较低，早期售后公房的建设密度则比较高。建设初期和如今的城市规划差距大，活动空间也比较局促，所以，相关配套设施随着生活水平的改善，更新需求日趋增加，这也促使街道开始了社区更新的工作。

早期城市发展和建设速度迅猛，几乎没有留白的城市空间。而老旧社区更新困难的根本原因是基础条件差，仅仅靠维修基金破局是很困难的。空间艺术季的启动对街道形成反向推力，对小区的社区生态造成很大的影响，社区可以先由细胞组成小的生态系统，再组合成更大的生态系统。

制度层面，在《物权法》《微更新管理条例》等已出台的政策之外，更新中也会触碰到规划上的问题，还需要进一步突破。这就需要业主委员会、居民委员会的支持，让渡一部分原本的社区空间，比如物业用房、社区单位的公共绿地等，以谋取更大的公共效益。2016年长宁区推出了“居民小区建设”行动。

例如，某小区中有个半地下车库没有消防设施，车库上方有个陈旧的小型活动室，如果仅仅在现状基础上完成“表皮更新”，功能结构并没有转化。所以我们进一步探索，利用改扩建的方式将地下车库翻新，除了配备助动车充电设施，还引入社区养老、活动、阅览等内容。规划部门在项目批复的过程中给予我们非常大的指导和支持。

传统的专家听证会模式经常收效一般，北新泾更新项目中采取了哪些与之相对的针对性策略?

张生义：首先要摸清社区人口结构，商品房、售后公房的人口结构完全不同。住在老公房里的年轻人沟通起来比较容易，比较排斥的往往是老年人，所以沟通上要着眼于这部分人群。从实际来看，设计师的沟通效果往往不太理想。街道可以搭建培训平台，培养一定数量和资质的居委干部。有个案例是新泾七村的亭子，我们采取的策略是通过局部搭建，搭建后请居民使用一段时间，得到反馈后再看是否需要调整。结果是，最初的反对声也逐渐消失。“美丽楼道”行动也遇到过类似的障碍，所以我们拟定了一些规则，比如签订共同维护楼道的协议，包括不得装二道铁门，不得在公共区域乱扔杂物等。规则先行，建成以后也要自行管理，实施以后效果逐步显现。管理和维护是社区更新过程中的另一个大问题。

社区规划师和一般意义上的建筑师、规划师等有何差异?

张生义：一般意义上的设计师，其实和政府的角色差别比较大。经验表明，社区规划师更倾向于从事景观和装置艺术设计。在具体的工作过程中，首先要反复进行实地调研，深度了解社区后再提出深化方案，此外还要着重考虑整体社区环境。设计师有设计师的优势，但也有局限。受限于规划调整和多部门沟通的现实境况，设计师在做社区规划项目时，付出的实际成本很高。

INTERVIEW WITH CURATOR AND ZHANG SHENGYI

(Beixinjing Sub-district Office Director)

Micro-Regeneration, an Effective Way of Redistribute the Urban Spaces

There are 8 sub-exhibition venues in this year's Urban Space Art Season, and the presentation content is different. For the site project of Beixinjing, how did the curatorial team select the sites and decide what to show on the exhibition?

First of all, judging from the pre-conditions of urban regeneration, there is almost no new available land in Changning District, so the main job is to revitalize the stock update. Hence, compared with other districts in Shanghai, the pace of construction in Changning is relatively slow. And if we take the Colorful Community in Pudong New Area as comparison, then we'll find that the scale of community regeneration in Beixinjing is relatively small.

At first, we chose the main site in Xinjing Wucun. At that time, a carport construction plan was designed to form a semi-open fitness venue and not disturbing the residents. We hope that after the opening exhibition, people can be led back to their community, but the plan turned out to be rejected. Later we learned that there was idle space in the Folk Culture Center, but this type of architecture space limited the way to exhibit, so we decided to build an empty shell while remaining the original appearance of the exhibition hall. After the exhibition was over and the boards were removed, it would look like nothing has ever happened. Specifically to the exhibition content, in addition to the 22 exhibiting cases applied by the sub-district, this exhibition mainly extends two parts: one is the systematic renovation of Tianshan West Road, the other one is the shadow play with urban space as a discussion proposition.

Compared with traditional way of display with boards, shadow play as an art installation contains the discussion of life appearance in Beixinjing community, also makes an extension of city life in Shanghai. So which core elements in urban regeneration have been mainly reflected?

This shadow play's name is "Be careful of the slippy street". The reason to choose this shadow play is that we happened to see the actors holding the puppets for shadow play on the exhibition site, so we tried to write some short films about urban space, street space and city life. Some of the content has been hidden from the script, such as dismantling. We also invited some professional actors to interpret the city life and played it repeatedly in video in the exhibition hall. We have created the process of exhibiting during the construction of the exhibition site, which can be regarded as the "'renovation exhibition' in the regeneration". It is a three-step exhibition, which is to educate based on regeneration and reflect the core of city life through the dialogues between the characters in shadow play, instead of simply "decorating" the interior space.

Although the systematic renewal project of Tianshan West Road wasn't implemented, the regeneration plan of road system is to be exhibited on this site project, which can give feedback to the local government. The district leaders can break through their previous perception after seeing the exhibition content, for example that the turning radius of the road can be reduced, and the road width can be narrowed. Specifying to the design, there were 6 car lanes on Tianshan West Road (including 2 sidewalks, 2 bicycle lanes and median strip), and we narrowed the main lane of 3.75 meters wide to 3 meters wide, releasing less than 3 meters of space for walking. Therefore, the street renewal is not to beautify the road, but to explore new potentials, such as the ability to turn three lanes into two lanes. Of course, this is related to the transportation mode advocated by the city. If the traditional lane can be changed into a more economic lane under the existing conditions, and the new space can be moved, then the width of the sidewalk will be increased by two meters to conduct more urban space renewal. It can also increase the numbers of lanes to resolve the conflicts caused by the concurrence of bicycles.

So I think the essence of urban regeneration is to release new space, not urban beautification. For residents, the internal renewal of the community addresses problems as big as urban open spaces, traffic space, the function of the entrance to the community, and as small as the lack of community open space, the greening and so on. These are the core issues.

On the curatorial result, is there a breakthrough in this site project?

During the project, we discussed non-motorized traffic system in the main exhibition, and explored the way to regenerate the community during the closing ceremony. So far, the content of the discussion will have discourse values for the professionals, but ordinary residents might not care much. So we hope to make some changes if possible, such as popularizing expertise to the general public, or trying to challenge something and asking questions about the city we live in. And we also hope participants to stand on a common ground and concentrate on amplifying the possibilities of one particular curation, rather than considering trivial renewal aspects. This goes back to urban system regeneration mentioned in the previous question, which means that we can't talk about tiny spaces while leaving the bigger system aside. Curatorial practice sometimes needs to break boundaries. For example, the West Bund Art Center was pushed to the forefront of space legalization during the exhibition period, and this status quo was able to change because of an Urban Space Art Season.

Reshaping the Cell of the Community's Eco-System

What is the root cause of the micro-regeneration of Beixinjing community?

Beixinjing sub-district used to be an area of urban-rural integration, 80% of the houses there were built in early 1980s. The proportion of commercial housing in the whole community is relatively low, and the construction density of early public housing is also relatively high. There is a large gap between the rationality of urban planning in the initial stage of construction and the current urban planning. Therefore, with the improvement of living standards, the related supporting facilities face more demand for renewal, and the activity space in the community is much more cramped. These actual needs prompted the street to begin to work on community regeneration.

At the institutional level, what breakthroughs have Beixinjing made?

At the institutional level, with the introduction of Real Right Law, Micro-Regeneration Management Regulations and other policies, some problems about planning will be involved during the process of regeneration, as well as some breakthroughs. Therefore, the support from the owners' committee and the residential committee is necessary to transform some of the original community space, such as buildings for property management, public green space in community etc. Last year, Changning launched "neighborhood building". There was a semi-underground garage in the community without fire-control facilities and there was a small old activity room above the garage. If the "facade renewal" is completed only on the basis of the status quo, the functional structure will not be transformed. Hence, we made further explorations to renovate the underground garage by means of reconstruction and expansion. In addition to providing the charging facilities for the locomotives, we also introduced community nursing, events, and readings. The planning department gave us great support during the project approval process.

The rapid urban development and construction in the early days led to almost no urban space left. The fundamental reason for the difficulty in regenerating the old community is the poor basic conditions, and it is very difficult to resolve the problem by the maintenance fund alone. The start of the space art season can have a great impact on the community ecology, forming a reverse thrust on the street. The community can first form a small ecosystem by cells and then combine them into a larger ecosystem.

Traditional community planning work may take a hearing. Generally, experts or influential people are invited to explain and communicate with the community residents. There may be many objections. What strategies have you taken in charge of the regeneration project?

First of all, we must find out the demographic structure of the community. The demographic structure of commercial housing and after-sale public housing is completely different. For young people living in public houses, communication is easier, and the older ones are often excluded. Therefore, communication should focus on this part of people, because from a practical point of view, the communication effect of designers is often not ideal. The street can build a training platform to train a certain number of qualified cadres.

There is a case of the pavilion in Xinjing Qicun. The strategy we took is to build partly, and invite the residents to use for a while, and then to see if there needs to be some adjustment after we get some feedback. By this way, the initial opposition turned into understanding. Beautiful Corridors had also encountered some obstacles. So we drew up some rules, such as signing an agreement to jointly maintain the corridor: First, two iron gates are not allowed to be installed, and throwing litter in public areas is also not permitted. Rules go first, and then residents should learn to manage themselves. Management and maintenance are also the biggest issues in community regeneration.

In your opinion, what's the difference between community planners and architects, urban planners and landscape designers? What rules should community planners obey during practical work?

Architects are quite different from the role of the government. And I think that community planners may be more inclined to do some landscape and installation work. First, we must conduct on-the-spot investigations, deepen understanding of the community and then propose a thorough plan, and we must focus on the overall community environment. However, designers have their own dilemma. Limited by planning adjustments, communication and coordination among various departments, the actual cost of doing community planning projects is very high.

展览基本信息

主题：新泾 · 新境

时间：2017.10.27—2018.1.12

举办地址：长宁区民俗文化中心南楼一楼、二楼（长宁区北渔路95号）

主办单位：长宁区人民政府

承办单位：长宁区规划和土地管理局、长宁区北新泾街道

协办单位：长宁区建设与交通委员会、长宁区文化局

策展人介绍

张佳晶

上海高目建筑设计咨询有限公司创始人、主持建筑师，上海交通大学客座讲师，上海市规划委员会专家。曾担任上海当代艺术博物馆"蜃景——当代中国博物馆建筑的十二种呈现"、徐汇西岸目外工作室"知微见著"等展览策展人，参与2012米兰三年展、2012上海当代艺术馆"生活演习"建筑空间艺术展、2011深港城市/建筑双城双年展"公共外延"等展览。

Basic Information

Theme: Xinjing, New Scape

Time: 27 October 2017 – 12 January 2018

Venue: 1st Floor & 2nd Floor, South Building, Changning Folk Culture Center (No. 95 Beiyu Road, Changning District)

Sponsor: Changning Government

Organizers: Changning Urban Planning & Land Authority, Beixinjing Sub-district

Supporters: Changning Construction and Transportation Commission, Bureau of Cultural Affairs

Curators

ZHANG Jiajing

Founder and principal architect of Atelier GOM, Shanghai; Guest lecturer at Shanghai Jiao Tong University; Expert of Shanghai Committee of Urban Planner. Curator of "SPECTACLE–12 Presentations Of Contemporary Museum Architecture in China" in Shanghai Contemporary Art Museum, and "HOLE TO WHOLE" by Atelier Moa at ART WEST BUND. Participated in LA THIENNAL DI MILANO (2012), "MOCK UP: ARCHITECTURE OF SPATIAL ART" at

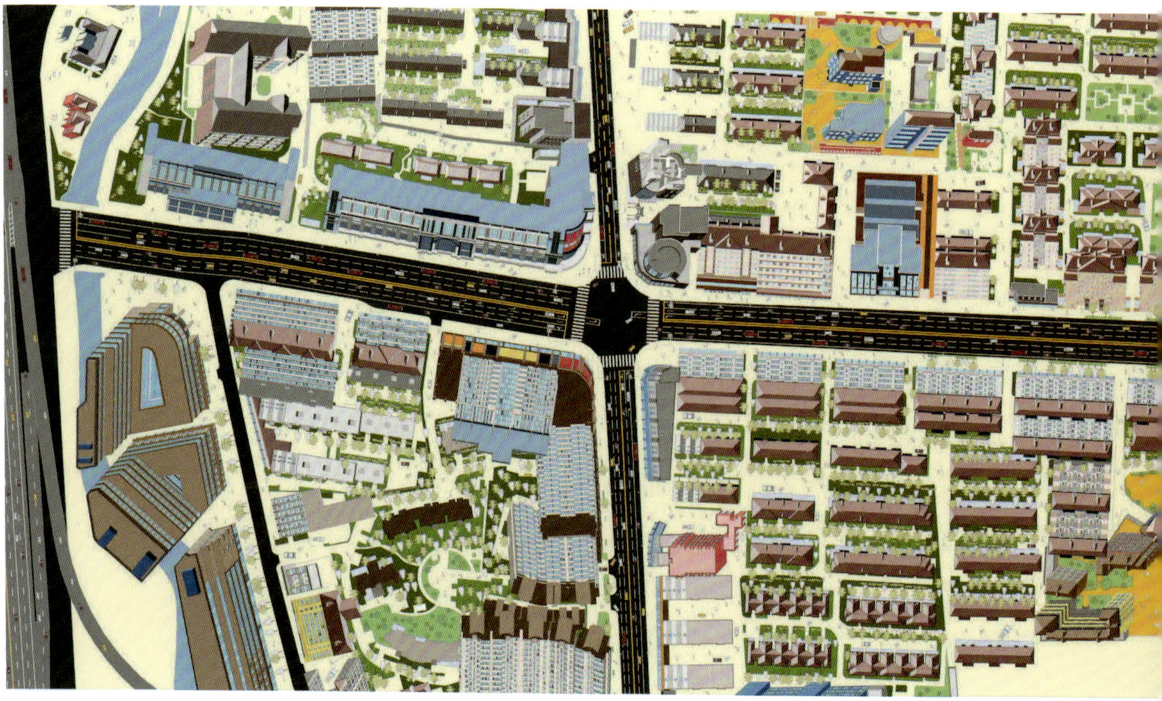

北新泾社区微更新长轴
Axis Picture of Beixinjing Community's Renovation

周渐佳

上海冶是建筑设计有限公司创始人、主持建筑师。同济大学客座讲师，香港大学上海中心客座讲师。曾参与包括2015四平街道创生展Open Your Space，2015上海城市空间艺术季，2013深港城市\建筑双城双年展，2012纽约设计周“NYC x Design：自然／建筑”展，2011成都双年展“物我之境／国际建筑展”等国内外重要展览。

采访+编辑

陈晨[城市中国]+崔国[城市中国]

图片提供

张佳晶、周渐佳

Museum of Contemporary Art (2012), "EDGE OF PUBLIC" at Hong Kong & Shenzhen Bi-City Biennale of Urbanism\Architecture (UABB)(2011) and other important exhibitions.

ZHOU Jianjia:

Founder and principal architect of Yearch Architecture Studio; Guest Lecturer at Tongji University, Shanghai and HKU/Shanghai Study Centre. Participated in "OPEN YOUR SPACE: Design Interventions in Siping Community" (2015), Shanghai Urban Space Art Season (2015), Hong Kong & Shenzhen Bi-City Biennale of Urbanism\ Architecture (UABB) (2013), "NYC x DESIGN Festival: Nature/Architecture" (2012), "HOLISTIC REALM" Chengdu Biennale (2011) and other important exhibitions at home and abroad.

Interview + Edit

CHEN Chen [Urban China] + CUI Guo [Urban China]

Image Source

ZHANG Jiajing, ZHOU Jianjia

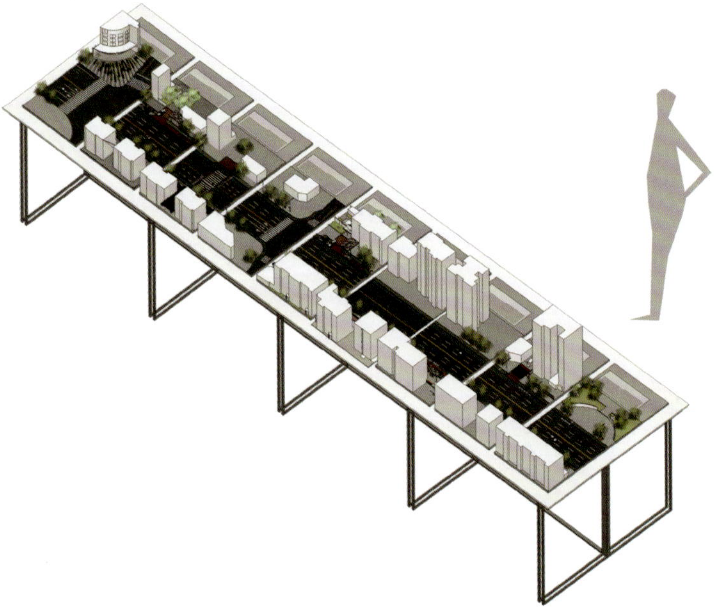

天山西路系统更新模型展台
Street model of Tianshan West Road after urban regeneration

皮影戏《侬当心伐要拐高》
The shadow play of Be careful of the slippy street

"更新中的'更新展'"，本次展览试图表现在施工现场进行展览的状态
Community renovation is creating the construction process

系统更新下的天山西路北渔路路口（未实施）
Renovated crossing of Beiyu Road at Tianshan West Road (unrealized)

改造后的天山西路步行道（未实施）
Pedestrian street after the renovation (unrealized)

设计师反复调研新泾五村，在不进行大改动的前提下，通过置入微形装置，引导社区居民，提升生活的便利性
The designer repeatedly made investigations and researches on the Xinjing Wucun and guided the residents to improve the convenience of life by placing micro-devices instead of making major changes

改造后的天山西路人行横道（未实施）
Crossing road after the renovation (unrealized)

历史与当下，重新焕发
生命活力的历史风貌空间

FROM THEN TO NOW: REVITALIZING
HISTORIC AREAS AND BUILDINGS

「符号上海」· 上海风貌保护道路文化之旅 · 上海城市规划展示馆实践案例展

2017 "ICONIC SHANGHAI": SERIAL EXHIBITIONS AT SHANGHAI URBAN PLANNING EXHIBITION CENTER

策展主题：符号上海

EXHIBITION THEME: ICONIC SHANGHAI

项目简介

"符号上海"作为上海城市规划展示馆聚焦上海城市精神提炼、城市文脉传承的城市文化展示品牌，创新地将规划专业展览与艺术创作和公众参与活动相结合，采用市民众筹、众议、众学的展览模式，打造了一个属于市民的城市文化舞台，得到社会各界的一致好评。如2015年的"上海石库门文化之旅"，将阮仪三教授的研究成果和连环画泰斗贺友直、"石库门先生"李守白的艺术作品相结合，并引入同济大学师生的里弄调查成果，使人耳目一新。2016年策划的"当艺术遇见邬达克"，在1000平方米的展厅中汇聚了将近20种展示（艺术）形式、近30个展项，近百位参展人作品，汇聚成一次彰显上海城市人文精神，具有专业性、新颖性和思想性的城市主题展览，并成功将"邬达克建筑"塑造成为时下"网红"。

策展主题阐释

城市符号是一座城市文化的缩影，是城市精神潜移默化的传承与延续，是定格在城市人心中永恒的记忆。有着七百多年建城史的上海，拥有丰富的文化符号，它们有机地构成了完整的沪派文化，是城市发展不可或缺的精神内涵。

展示内容

2017年恰逢上海市政府颁布《关于本市风貌保护道路（街巷）规划管理的若干意见》，确定64条"永不拓宽"道路10周年。为此"符号上海"决定聚焦"街巷"这个汇聚城市文脉个性、展现城市活力风韵、体现城市发展变迁、与市民生活日常休戚相关的主题，并延续以往几届展览"以城市精神为核心议题、以保护案例树发展理念、以艺术创作促当代思考、以市民作品展城市风采、以公众参与聚城市活力"的品牌特色，进一步提升展览创意的多样性，展现不同视角和不同人群对城市历史文化和风貌遗产保护的认知与努力。

PROJECT INTRODUCTION

Hosted by Shanghai Urban Planning Exhibition Center (SUPEC) as an urban branding exhibit project, "Iconic Shanghai" sheds light on preserving the city's urban spirit and revitalizing its cultural heritages. It has created a stage for civic culture by adopting the multi-participation model of "crowd funding, public discussion, and mass learning". The serial events creatively combined curated exhibition, artwork creation and public interaction, and were thus positively received. A refreshing example is the Shanghai Shikumen Cultural Tour in 2015 that brought the research of Prof. Ruan Yisan, works of famous cartoonist He Youzhi and "Mr. Shikumen" Li Shoubai, as well as survey reports about Lilong lanes conducted by teachers and students of Tongji University. In a 1,000m^2 venue, "When Art Meets Hudec" (2016) employed nearly 20 forms of arts, roughly 30 items, and works created by almost a hundred artists to offer a professional, innovative, thoughtful exhibition that embodied the humanistic spirit of Shanghai. László Ede Hudec, a Hungarian who worked in Shanghai during 1920s and 1930s, with his architectural heritage in Shanghai, became viral on Chinese internet.

THEME INTERPRETATION

As never-fading memories of its residents, icons epitomize, inherit and deliver the culture of a city. Shanghai, with its history of over seven centuries, boasts a variety of symbols, which constitute an organic part of Shanghainese culture and the indispensable spiritual backbone for its progress.

EXHIBITION

In 2017, Shanghai Municipal People's Government issued *Some Suggestions on the Administration of Planning Work for Preservation-of-Historical-Look Streets (Alleys/Lanes) in This Municipality*; ten years ago, the list of "64 expansion-prohibited streets" was published. Echoing these events, "Iconic Shanghai 2017" focused on "Streets" as a theme, which is an element that embodies the character, energy, history, and lifestyle of this city. Carrying on the established feature of "Taking the spirit of Shanghai as central discourse, Corroborating development concept with preservation cases, Promoting contemporary thinking with art, Demonstrating the city's charm with works of its residents, Enhancing the city's vitality with public participation", Iconic Shanghai 2017 emphasizes diversity by presenting different perspectives and endeavors of various groups regarding the city's history and preservation.preservation.

相较于之前的城市总体规划，上海2035总体规划中最鲜明的特点之一是以“提升城市发展品质”为目标引领。这一转变一方面是为了回应老龄化程度日趋严重，人口结构、文化观念更加多元的未来社会，另一方面更体现了政府在城市建设中将目光逐渐从空间区块等产业分布要求转移到市民认同、需要和希望的“对美好生活向往”上的努力。总体规划强调了建设人文之城的四个努力方向，分别为：构建多元融合的15分钟社区生活圈、保护风格独特的历史遗产、塑造特色凸显的城乡风貌、培养兼收并蓄的文化氛围。其中，15分钟生活圈以社区为单位，强调居民日常生活中公共空间、交通、服务设施的便捷可达；保护历史遗产从历史文化要素入手，关注文化保护对象及其保护机制的建立；特色凸显的城乡风貌从小空间看大格局，着重强调只有积累一定数量富有人性关怀的公共空间才足以塑造富有人文特色的城乡风貌；兼收并蓄的文化氛围区别于前三个偏重具体空间的措施方向，更强调一种抽象的城市氛围与气质特征，而氛围与空间塑造的基础都离不开对人的重视，对市民文化素质的培养与提升，市民的精神面貌与文化审美是城市文化氛围的重要表现之一。

街道作为城市中数量最多、分布最广、密度最高的公共开放空间，不可避免地承担了城市形象重要窗口这一现实任务。游客通过街道来认识一座城市，身处其中的市民也通过街道加深对这座城市的理解和认同。横纵交错的街道、弄堂在物理层面将城市串联起来，又在文化层面传承了属于一座城市的历史脉络。有着七百多年建城历史的上海，将海派文化的基底深深烙印于市民内心并获得认同，这不仅是践行总体规划以人为本理念的完美体现，更是推动上海比肩其他全球城市的重要保证。

城市符号诠释城市文化

由于是政府颁发的专业性政策文件，城市总规虽然有专门的公众读本以拉近和普通市民的距离，但不可避免地会遇到受众窄、解读难等一系列问题。城市的发展愿景单纯靠政府、专业学者及部分参与总规修编的公众代表的支持是难以走得长远的，如何将总规理念以切实可感、通俗易懂的方式传达给市民大众，一直是规划资源等政府部门关注和努力的重点。上海城市空间艺术季的案例展之一——“符号上海”上海风貌保护道路文化之旅系列展览活动，就是上海城市规划展示馆向大众传播城市规划科学价值观、解构城市精神文化内涵、增强受众对城市地域文明和时代文化认同的重要举措之一。

“符号上海”是上海城市规划展示馆长期坚持的城市文化品牌推广项目，过往已举行过多次主题展览，均是城市文化大主题下的具体实践。而本次城市空间艺术季实践案例展的重心则放在上海风貌保护道路这一主题上，从“街道”这一最重要又最容易被人忽视的城市公共空间入手，以小见大，通过街道的风貌演变重新唤醒大众对城市文化的重视与期待。城市文化品牌的塑造大致要经历以下几个阶段：首先是文化觉醒，即市民知晓自己所居住的城市有怎样的文化背景；其次是文化认同，在清楚自己城市文化背景的基础上，尊重、认同其文化底蕴并认识到这样的文化品牌对城市发展的重要性；最后才是市民在一定程度文化认同的基础上，对城市文化品牌的参与和塑造。“符号上海”一系列主题展览活动就是想通过这些丰富生动的活动激发市民的文化觉醒，营造公众参与城市发展的“对话”氛围，为上海2035美好未来的实践，注入来自公众的呼应。具体来说，为达成案例展的有机性、丰富性与多样性，“符号上海”采用室内展览、室外展览、公众活动等多种类型，分为规划馆系列展览、户外集装箱展示活动、市民竞赛活动和主题系列讲座四大部分，展现了不同视角和不同人群以街道为切入点，对城市历史文化和风貌遗产保护的认知和努力。其中，“悦读城市”系列讲座和海派手工技艺传习所偏重面对面的交流，让市民在听和动手的过程中感受文化魅力；户外展示项目、市民摄影和绘画比赛均是希望通过一种大范围的交流形式吸引更多市民参与；规划馆系列展览则融前三部分之成果，佐以丰富的补充内容，用照片、画作、艺术装置等实物为大众带去关于街道、城市的心灵碰撞。规划馆系列展览共有八个，分别针对街巷的记忆、街巷的风情、街巷的彩韵、街巷的童趣、街巷的空间、街巷的经纬、街巷的城乡对比以及街巷的中外对比。

时空回眸“溯”文化烙印

“街巷• 记忆”通过新老照片对比，街道历史谱系、历史图纸和影像资料的挖掘，剪纸（装置）等艺术作品的全新创作，配合系列讲座、市民课堂、城市定向赛、市民摄影赛等市民参与活动，在力求全面地反映上海道路（街巷）文脉和发展进程的同时，用传承的理念、艺术的语言、专业的讲座、公众的参与，引导市民走进展馆、走进街道，加深市民对上海城市风貌保护的认知与尊重。展览由“溯源”“探秘”“保护”“传承”四个板块构成。“溯源”板块，以场景营造和历史解读将64条“永不拓宽”的风貌保护道路从历史、文化、风貌、生活等视角进行系统剖析，在向观众展现这64条风貌道路保护价值的同时，也表达了向10年前风貌道路保护规划的编制者们的致敬之情。

“探秘”板块，全面介绍了64 条“永不拓宽”道路的名称、区域、发展概述、风貌特色、历史故事，并通过新老照片对比、64件针对每条道路创作的艺术作品、影像资料等综合展示方法，提升展览的艺术性和可读性；“保护”板块，以上海风貌道路保护工作概况介绍、徐汇区风貌道路保护案例、新场古镇的老街保护案例等上海不同特质街道的保护现状与成果等丰富内容展现上海在风貌保护道路工作中的理念与成就；“传承”板块，由街巷“悦读”（书吧）、海派手工技艺传习所、科普游戏、风貌保护道路旅行地图、心愿墙等活态的展项组成，体现了上海市民对风貌道路保护的参与热情。

街巷经纬“刻”历史变迁

“街巷•经纬”由上海城市规划展示馆和上海市测绘院共同主办。地图是一座城市发展变迁的客观记录者，具有丰富的历史底蕴与人文内涵，历史地图如同城市成长的年轮，记录着城市街巷的演变、城市文化的传承，散发着这座城市独特的内在魅力和气质。展览通过新老地图的演变与更替，以城市街巷为主题元素，以“缘起与演变”“留影与印鉴”“传承与更新”三个版块内容，以地图经纬线的方式生动再现城市街巷的历史变迁与文化印迹，呈现了城市空间演进对上海这座国际都市的浓厚文化内涵形成的深远影响。

文学作品“悦”城市风情

“街巷•风情”是国内首次以文学作品为创意核心的艺术展。策划灵感来源于著名作家陈丹燕撰写的一本属于上海的文学作品《永不拓宽的街道》，整个展览凸显着上海独有的都市风情，清新而文艺，典雅而富有内涵。展览由19篇《永不拓宽的街道》书摘构成的“认识上海”“街道印象”“生活日常”“他者之眼”“理想实录”五个叙事空间串连起观展脉络。王劼音、方世聪、陈逸鸣、俞晓夫、黄阿忠等江、浙、沪顶级油画大师的64幅作品，将上海风貌保护道路的历史风韵展现得淋漓尽致。

以书策展、展蕴书香。展览巧妙地利用空间重塑和场景营造，使作家、画家、市民、观众都成为展览的一部分，创造了一次与众不同的文化悦读体验。

非遗技艺“绘”城乡交融

“街巷• 彩韵”由上海城市规划展示馆与金山区文广局联合主办。展览以独具特色的金山农民画为载体，用农民艺术家的独特视角，勾勒出农村居民眼中的城市百态。此次参展作品是金山农民画艺术家首次面向城市，用金山农民画派的“非遗”技法，全面、整体地反映城市题材的创新实践，这既是城乡绘画艺术的互鉴，更是城乡文化观念的交流和碰撞。为城市艺术的多样性注入来自田园的清新与活力。金山农民画是滋生于江南民间艺术土壤的传统技艺，通过造型稚拙、色彩强烈、构图饱满、风韵独特的画面，展现出江南水乡特有的乡土人情，散发着浓郁的地方生活气息。作为金山区靓丽的文化名片，金山农民画带着泥土的芬芳，根植于金山的乡土之中，孕育在金山农民艺术家之间。通过40年的培育和发展，金山农民画逐步走向全国，走出国门，成为上海的特色文化符号之一。

联合主办。展览为同济大学建筑与城市规划学院教授、上海美术家协会城市环境与壁画艺术工作委员会副主任阴佳先生的个人作品展。展览既是对阴佳先生三十年艺术创作与教学生涯过程中，行走城市乡间、游历世界各地后的感悟与创作的回顾，更是对所有热爱城市、坚持城市题材艺术创作的城市艺术家的尊重与褒奖。展览汇聚了阴佳先生三十年来创作的油画、水粉、素写、版画、浮雕、壁画、装置、雕塑等诸多艺术领域的优秀作品。这些不同类型、工艺、意境的艺术作品，不仅体现城市与乡村的个性、空间与个体的关联、传统与现代的交融，更从侧面反映了三十年来城乡空间与面貌的时代变迁。

童谣童趣“梦”美好未来

上海童谣和弄堂游戏伴随着一代又一代上海人的成长，已成为上海人的记忆传承和上海独具地域特色的文化符号。“街巷·童趣”展览在上海市工艺美术大师李守白先生创作的《上海童谣》剪纸长卷的映衬下，向观众呈现了由上海城市规划展示馆与上海市青少年活动中心联合发起组织的第二季“娃娃画上海”儿童绘画作品大赛的优秀作品。绘画大赛的主题是“我的街巷，我的家”，参赛儿童围绕自己身边的街道、景观、建筑等城市风貌进行主题创作，展示作品充满了童趣童真，所描绘的城市空间更是展现出儿童特有的想象力和创造力。展览活动的举办，在引导孩子们走近城市、感受生活、表达心愿的同时，也让成人朋友们有机会认识孩子们关于城市发展的期待与梦想，有利于为下一代创造更为安全、美丽、文明的城市，使城市空间艺术具有生命力，具有传承性，具有持续发展的品质和前景。

中西互鉴“育”多维视角

“上海•威尼斯城市空间之旅”由上海城市规划展示馆、同济大学建筑与城市规划学院、威尼斯建筑大学环境规划与设计学院联合主办。该展览是对同济大学建筑与城市规划学院师生围绕上海与威尼斯两个东西方城市的水岸空间开展的艺术写生与创作作品的集中展示。“水岸色韵——上海•威尼斯城市空间之旅”展览由“城市记忆”“城市欲望”“城市符号”“城市色彩”四个板块构成，展示了22位师生创作的近200件作品。“城市记忆”着重展现对城市过往的记录，如威尼斯大运河和上海外滩的立面长卷；“城市欲望”则描绘城市为居民生活所营造的“秀场”，如威尼斯狂欢节的人群、上海徐汇滨江的市民活动；“城市符号”聚焦于城市独特的文化样式，如威尼斯府邸建筑常用的窗洞形式，上海石库门的典型样式等；“城市色彩”则展示不同学生对同一城市空间或同一主题的不同视角的认知与思考。

画境之外“展”迁想妙得

“画境之外——城市与乡间的时空印记”由上海城市规划展示馆与同济大学建筑与城市规划学院

馆校连接“促”开放思维

由上海城市规划展示馆与同济大学建筑与城市规划学院联合主办的“连接——城市公共空间的多元视角”展示了大学教育与研究课程中对城市人

文的关注与重构。展览将同济大学建筑与城市规划学院“2017 TAP 联合设计工作坊”“当代装置艺术工作坊”“当代影像艺术实践与解析”的教学与研究课程，与针对上海地域性城市公共空间（以城市街道为例）的空间模度研究相结合，课研的最终成果以展览的方式向公众呈现，促进高校课研成果的社会共享转化，展现高校教学科研的国际化、创新性和多元融合，并以此激发市民的思考，引导市民更为深切地关注城市街道公共空间的保护与更新。展览由艺术装置（基于所调研的上海街道空间与生活意象进行设计）、图文解读（针对装置作品所对应街道的空间模度与公共生活活力研究和公共空间设计策略）、街道调查实录（照片、手绘、影像）组成，结合展厅环境氛围的营造，在体现教学成果展的专业性的同时，融入工作坊的开放与活力思维。

Compared to the previous Shanghai Master Plan, one of the most striking features of the new edition (2017-2035) is its objective towards "quality urban development". On one hand, this change is a response to the aging society and the diversifying population and cultural ideology. On the other hand, this manifests that the government has gradually shifted its vision from the spatial distribution of urban industries to the citizens' "hopes and needs in urban life". In order to build a humanistic city, there are four promoted approaches: to build diverse and inclusive community living circles in 15 minutes' walks, to conserve unique historical heritages, creating featured urban-rural landscapes, and cultivating an inclusive culture milieu. Each 15-minute living circle would cover a community, focusing on improving the accessibility of public space, transport and other facilities within walking distance. Secondly, historical heritage preservation pays special attention to establishing a systematic protection scheme based on keen historical and cultural features. The third approach, by reading the big picture from a small scope, addresses the importance of humanistic public spaces in shaping the urban and rural environment with rich cultural characteristics. Last but not least, the fourth approach differs from the other three by paying more attention to the atmosphere and impression of the city rather than physical space. As the shaping of a city's atmosphere and space is inseparable from its dwellers, without improving public aesthetics through education, it would be hard to achieve such cultural atmosphere. Streets, as the most common open spaces in the city, are important images of the city. Tourists come to know a city through streets, and citizens gain understanding and recog-

nition through streets. The network of streets and Lilong connects the cities physically, while also links its historical context in cultural terms. For Shanghai, a city with more than 700 years' built history, to deeply imprint the essence of Shanghainese culture on people's memory is not only an incentive of the Master Plan, but also an important move to promote Shanghai as an excellent global city.

A SYMBOL FOR URBAN CULTURE

The Master Plan is a professional policy document issued by the government. Although a more legible edition has been published to close the gap between the government and the public, it also inevitably encounters problems such as low comprehensibility and limited audience. As a city's long-term vision could not solely rely on the effort of government, scholars and public representatives, to convey the ideas of the Master Plan to a wider public in a practical and understandable way has become the mission of government departments such as the Shanghai Urban Planning and Natural Resource Administrative Bureau. Therefore, one of the site projects at Shanghai Urban Space Art Season - "Iconic Shanghai" serial exhibition, is one of the most important acts for SUPEC to, towards the public, showcase the methodology of urban design, transmit urban cultural values and enhance the cultural identity of citizens.

"Iconic Shanghai" is an existing cultural brand of SUPEC. There have been several exhibitions held under similar themes, all of which are specific practices concerning urban culture. The focus of this site project is conservation within Shanghai's unique style. "Street", as one of the most important and easily overlooked urban public spaces, may be the most potential element to reactivate the public's attention toward urban culture. The formation of an urban cultural brand goes through the following stages: first, cultural awakening, that is, citizens can understand the cultural background of the city; secondly, cultural identity–on the basis of comprehending urban culture–citizens could identify their cultural heritage; the last step is to encourage public participation in the brand building upon a certain degree of cultural recognition. The "Iconic Shanghai" exhibition series is an attempt to bring awareness and discussions of local culture through fruitful activities, and to embed a humanistic, public-engaging setting for the practice of "Shanghai 2035".

Specifically, in order create an organic and diverse case exhibition, "Iconic Shanghai" adopts various forms of exhibition-indoor display, outdoor exhibition, and interactive activities. They are divided into four parts: serial exhibitions in the museums, outdoor container displays, competitions among local residents and lecture series, each showing different groups' perspectives towards urban cultural heritage conservation through the idea of the street. Among them, "Grateful City Readers" lecture series and workshops on Shanghainese craftsmanship are face-to-face activities that enables the public to experience culture through listening and making. Outdoor display, photography and painting competition are oriented toward a wider public. On the other hand, the exhibition series in the Urban Planning Exhibition Center yields the fruits of three other parts, bringing sentimental touch about streets and the city through rich photographs, paintings, installations and other media. There are eight exhibitions in the Planning Museum, which presented the memory of streets, the style of streets, the color of streets, the childish joy of the streets, the spatiality of streets, the latitude and longitude of streets, the urban-rural contrast of streets, and the comparison between Chinese streets and streets worldwide.

TRACING CULTURAL IMPRINTS THROUGH TIME

The "Streets • Memory" exhibition presented comparisons of historical and new photographs, historical pedigree of streets, old planning drawings and images. It also organized activities such as paper cutting, lectures, orienteering, photography competitions and other public events. All had aimed to reflect the background and development of streets in Shanghai, while also attract citizens to get involved, walk into the streets and gain stronger apprehension. The exhibition consists of four sections: Tracing, Exploring, Protecting, and Inheriting. The "Tracing" section introduced the history, culture, image, and lifestyle of the "64 expansion-prohibited streets" with scenes and documents, acknowledging efforts made by policy makers in the previous Master Plan. The "Exploration" section delineated the name, location, development overview, features, historical stories of each street through images, art works and video recordings, creating a both artistic and legible narrative. The "Preservation" section focused on the concepts and achievements in Shanghai's streets preservation, sharing case studies such as the featured road renovation projects in Xuhui and Xinchang old town. The "Inheriting" section portrayed the passionate interactions between streets and people, composed of events such as the reading gallery, craft workshops, popular science games, art map of conserved roads, and the wall of wishes.

HISTORY MARKINGS IN-BETWEEN LATITUDES AND LONGITUDES

The "Street • Latitude & Longitude" exhibition was co-organized by SUPEC and Shanghai Surveying and Mapping Institute. Maps are the objective urban observers with historical depth and humanistic connotation. Maps to a city are like annual rings to a tree, marking the evolution of streets and the bearings of urban culture, depicting its unique charisma. The exhibition was composed of three sections, "origin & evolution", "impression & imprint" and "inheritance & renovation" under the theme of urban streets. In-between latitudes and longitudes, the historical markings and cultural imprints on the maps implicate the role of urban space making in shaping Shanghai's profound urban culture as a global city.

READING URBAN CHARM THROUGH LITERATURE

The "Street • Styles" exhibition, the first art exhibition in China with literary as its source of creativity, was inspired by Streets That Were Never Expanded, a book about Shanghai by famous writer Chen Danyan. The exhibition highlights Shanghai's unique urban style that is fresh and artistic, elegant and sophisticated. 19 quotations from the book shape the narratives of the exhibition, in the form of five connected spaces with the subtitles of "Knowing Shanghai", "Street Impression", "Daily Norms", "Alternative Stories" and "Documenting Ideologies" respectively. 64 artworks from renowned painters of the Yangtze River Delta region, such as Wang Jieyin, Fang Shicong, Chen Yiming, Yu Xiaofu, and Huang Azhong, vividly depicted the historical charm of the Shanghai featured streets. The exhibition was curated with and accented by the book, creatively using spaces and scenes to make the author, the painters, and audiences mingle in a unique city reading experience.

PORTRAYING URBAN-RURAL EXCHANGE WITH NON-MATERIAL LEGACY

The "Street•Color" exhibition was jointly organized

by SUPEC and Shanghai Jinshan Administration of Culture, Radio, Film & TV. The exhibition used Jinshan peasant paintings to outline the urban life in the eyes of rural residents. It was the first time that Jinshan peasant painting artists used this non-material legacy technique to comprehensively present innovative practices of urban subjects. This is a process of not only skill exchange between urban and rural art, but also collision between urban and rural ethos, bringing freshness and vitality of the rural life into the diverse urban art. Jinshan peasant painting is a traditional art form rooted in the southern regions of the Yangtze River, characterized by simple figure, bold color, frame-filling composition and exceptional style. It reflects the unique local customs, folk culture and traditions of water villages south of the Yangtze River. As a distinctive cultural icon in Jinshan District, Jinshan peasant paintings are seeded in the native land of Jinshan and matured in the hands of local artists. Through 40 years of refinement, Jinshan peasant paintings have gradually become one of Shanghai's distinctive cultural symbols towards the world.

BEAUTIFUL FUTURE IN NURSERY RHYMES

The nursery rhymes and Lilong games of Shanghai, accompanying the growth of Shanghai people from generation to generation, have become Shanghai's cultural heritages and important symbols. The "Street • Childish Joy" exhibition, under the backdrop of the "Shanghai Nursery Rhyme" paper-cut scroll by Shanghai arts and crafts master Li Shoubai, presented the excellent work of the second "Children Painting Shanghai" competition. This exhibition was organized by SUPEC in cooperation with Shanghai Youth Activity Center. These drawings in the theme of "My Street, My Home" illustrated streets, landscapes and architectures in children's eyes, in an innocent, playful, imaginative and creative manner. Through the exhibition, children are encouraged to approach the city, feel the breath of urban life and express their own imagination. On the other hand, it also gives adults the opportunity to pay attention to children's visions and expectations of urban development, and thus shape a safe, beautiful and civil environment for the next generation. Moreover, it would create vitality and sustainability for urban space art.

REMARKABLE CONCEITS PRESENTED BEYOND DRAWINGS

The"Beyond Drawings---The Spatial and Temporal Markings in Urban and Rural" exhibition was the joint effort of SUPEC and the College of Architecture and Urban Planning of Tongji University(CAUP Tongji). It is a solo exhibition of Mr. Yin Jia, professor at CAUP Tongji and associate director of the Urban Environment and Mural Art Committee of the Shanghai Artists Association. The exhibition offered a retrospect into the insights and creations of Mr. Yin's in his 30 years' artistic and teaching career, in urban and rural areas around the world. This exhibition also demonstrated respect for all the urban artists who are passionate about and devoted to the artistic creation of the city. The exhibition brings together outstanding works in the forms of oil painting, watercolor, sketch, woodcut, relief, mural, installation, sculpture and so forth, in Mr. Yin's creative journey of 30 years. These works in different media and techniques not only reflected the character of the urban and the rural, depicted the connection between space and individuals, illustrated the convergence of tradition and modernity, but also manifested the transformation of the urban and rural areas in the past three decades.

MULTIPLE PERSPECTIVES BRED FROM INTER-CULTURAL LEARNING

The "Shanghai•Venice Urban Space Journey" exhibition was a collaboration between SUPEC, CAUP Tongji, and the Department of Design and Planning in Complex Environments of Università Iuav di Venezia. The exhibition displayed sketch drawings and creative works of the faculty and students of CAUP Tongji in the waterfront area in both Shanghai and Venice – two water cities in the East and West. The exhibition, "Waterfront and Rhyme – Shanghai•Venice City Space Journey", consists of four sections: "Urban Memory", "Urban Desire", "Urban Symbol", and "Urban Color", with nearly 200 works created by 22 teachers and students. The "Urban Memory" section focused on works that document the city's histories, such as elevation drawings of the Grand Canal in Venice and the Shanghai Bund. "Urban Desire" depicted vivid scenes of urban life, such as the crowd in Carnival of Venice and public activities along the Xuhui Riverside. "Urban Symbol" stresses on the unique formal languages of the cities, such as the types of window openings used in the façade of Venetian architecture, and the architecture types of Shikumen in Shanghai. "Urban Color" puts its eye on how different students perceive the same urban space or discourse through divergent lens.

AN OPEN MINDSET FROM ACADEMIC COLLABORATION

Brought together by SUPEC and CAUP Tongji, the "Street • Space" exhibition delineated the rethinking and reconstruction of urban humanities in the curriculum of higher education. The exhibition integrated the researches from three courses taught at CAUP Tongji-the "2017 TAP Joint Design Workshop", the "Contemporary Installation Art Workshop", and the "Contemporary Video Art Practice and Analysis"-into a research on the scale modules of Shanghai's public spaces(with streets as the paradigm). With these researches interpreted and presented to the public, the exhibition promoted knowledge sharing between academics and the society, while also demonstrating the institute's international vision towards innovation and integration. All these efforts would help evoke the public attention toward the conservation and renovation of streets. The exhibition consists of art installations (designed upon images of the streets from the research), graphic interpretation (about modular studies and researches on public life vitality of the installations, and the design strategies of the public spaces), and documentation of street surveys (in photographs, hand drawings and videos). Integrated with atmosphere of the exhibition hall, all these high-quality contents perfectly presented the professional teaching achievements of the college, in the meantime echoed the openness and vitality of the workshops.

"悦读城市——规划科普巡展"通过多媒体集装箱、移动展板、标准化可组装的矩形展箱、VR互动、科普影片、科普游戏等多种展示形式，帮助市民更深层次地了解上海、认知上海，感受城市文化对城市发展、城市生活乃至城市规划变革的深刻影响

"Pleasant Reading of City – Exhibition of Planning" helped the public to recognize, understand and experience more deeply about urban life and planning strategy in Shanghai through multimedia containers, mobile display panels, standardized rectangular display cases, VR interactive activities, science videos, science games and other exhibit forms

"街巷•活力"城市定向赛，让公众在运动中体验上海风貌保护道路的魅力
City Orienteering Tournament "Streets and Vitality" helped participants capture the glamour of streets in Shanghai

面向亲子家庭的海派文化手工体验课程，旨在让中小学生通过学习海派手工技艺，体验上海风情，感受海派文化
Shanghainese craftsmanship workshop for children, designed to enable school students to experience Shanghai culture by learning the skills of handwork

上海市工艺美术大师李守白先生创作的海派纸雕艺术作品。上：《永嘉路》；左下：《思南路》；右下：《太原路》，于"街巷•记忆"中展览。

Haipai style paper cutting works: Yongjia Road (above), Sinan Road (bottom left), Taiyuan Road (bottom right).The artist, Li Shoubai, is a master in arts and crafts, sculptures and paper cutting

“街巷•风情”展览凸显上海独有的都市风情，清新而文艺，典雅而富有内涵
The exhibition "Street • Styles" revealed Shanghai's unique vibe, incorporated with freshness and artistry, elegance and substance

金山农民画艺术家用农民画派的“非遗”技法，以激情的色彩和绚丽的画面描绘城市的街巷风貌。左上：怀明富《虹桥路》；左中：邱建国《思南路》；右上：邵其华《溧阳路》；右中：朱连君《雁荡路》；右下：陈卫雄《甜爱路》

Jinshan peasant painting artists use the unique techniques of peasant painting to depict the city's streetscape with passionate colors and gorgeous pictures. Upper left: "Hongqiao Road" (HUAI Mingfu); Middle left: "Sinan Lu" (QIU Jianguo); Upper right:"Liyang Road" (SHAO Qihua) ; Middle right: “ Yandang Road ” (ZHU Lianjun); Lower right: "Tian'ai Road" (CHEN Weixiong)

“街巷•童趣”展览以“我的街巷，我的家”为策展核心，通过“娃娃画上海”儿童绘画作品和“上海童谣”剪纸长卷两种艺术形式的组合，展现了不同时代的市民在同一个城市，拥有的不同童年经历

The exhibition "Street • Childishness" was curated around the theme of "My Streets, My Home". Through children's paintings from "Depicting Shanghai by Kids" competition and paper-cut long scrolls from "Shanghai Nursery Rhyme", it showcased different childhood experiences in different times

"画境之外——城市与乡间的时空印记"展览，既是艺术展，更是通过艺术的方式折射城市文明进程的观念展
"Beyond the Painting - Time and Space Imprint of Cities and the Countryside" was also a collection of reflective ideas on urban civilization

《上海童谣》剪纸长卷，代代相传的儿童童谣与惟妙惟肖的形象刻画
Work from "Shanghai Nursury Rhyme" paper-cut scroll, depicting children's rhymes and vivid images inherited from generation to generation

采访：策展人翁文斌

您策划此次案例展的初衷和感受是什么？

一座好的城市和一件好的艺术作品一样，不仅需要有可以让人驻足留恋的感动，更需要给予对方“对话”的空间、“对话”的语境，甚至“对话”的欲望与能力。所以城市规划展示馆存在的价值并不仅仅是以规划为议题构建起政府与市民之间的沟通桥梁，更是要通过发掘一个又一个公众能够感受得到的“触点”，制造一拨又一拨可以持续发酵的“故事”，以此激发公众形成对城市文化的探索热情，驱动社会形成对城市议题的追踪热情，营造在时代文化氛围下的市民与城市的“对话”空间，培育在科学规划思维下的市民与城市的“对话”语境，提升在时代文明思潮熏染下的市民与城市的“对话”能力，继而推动市民文明、社会文明、治理文明、城市文明的同步提升。这就是规划馆构建“符号上海”这一城市文化品牌活动的初衷，也必将是规划馆得以持续得到市民关注与追捧的关键。

城市的文脉传承与风貌保护并不是一个固化的名词，而是一个需要结合时代发展不断地反省、持续地再认知的过程。在不断的认知与反省中市民与城市两者的愿景才能更为趋同，才能有助于社会将对文化自信的热情转化为对文化传承与发扬的自觉。

街巷是连接城市与居所的物理空间，就像血脉一样对于城市这一生命体是必不可少的。街巷也是人们传承城市历史文化、感受城市精神与活力的重要载体，人们对于城市形象的认识和了解，往往是在对街巷空间形态与文化风貌的实地体验中逐步完成的。在承载着丰富历史文化信息的城市空间中行走，才能真正体会到城市的温度，感受到城市的活力。

INTERVIEW WITH CURATOR WENG WENBIN

May I ask your original intention and feelings after curating this case exhibition?

A nice city is like a piece of art: with not only sentimental value to make people linger, but also intellectual value to offer the context, space, and motivation for dialogues. Therefore, the incentive of the Shanghai Urban Planning Exhibition Center(SUPEC) is to not only bridge the government and the public on planning issues, but also discover the focal points where the public feels compassionate, take actions and spark their own stories. These series of actions would further evoke aspiration for urban culture and urban discourses among citizens, and thus create the space for dialogue between the city and the public within the contemporary culture and under the mindset of scientific planning. The civil life, social life, urban governance, and urbanity of a city could not be improved without putting efforts into shaping the citizens' ability to generate conversations. This was the original intention of SUPEC's "Symbolic Shanghai" urban cultural branding project, and it will undoubtedly continue to be the key for SUPEC to receive public attention.

The conservation of a city's cultural heritage and style is not an abstract concept, but a practical process of continuous reflection and interrogation in the context of time. Only through this can the city's vision converge that of the public, and turn the society's enthusiasm and cultural confidence into actions for revitalization and development.

Streets and lanes are the physical spaces that connect dwellings with the rest of the city, just like blood vessels to a body, they are essential to the life of the city. They are also important carriers of urban history, culture, spirit and vitality. People's understanding and recognition of the image of the city is gradually formed through intimately experiencing the spatiality and cultural atmosphere of the streets. Walking in the urban space with rich historical and cultural value is always the most direct way for one to feel and truly understand the vitality of the city.

展览基本信息

主题："符号上海"·上海风貌保护道路文化之旅系列主题展览活动

时间：2017年9月9日—2018年1月28日

举办地点：上海城市规划展示馆

主办单位：上海城市规划展示馆

协办单位：同济大学建筑与城市规划学院、上海市测绘院、上海市青少年活动中心、上海守白文化艺术有限公司、上海安墨吉建筑规划设计有限公司、上海交通大学城市更新保护创新研究中心、上海市金山农民画院、南京大学出版社、上海蒂比文化发展有限公司等

策展人介绍

翁文斌

上海城市规划展示馆总工程师。2000年1月入职上海城市规划展示馆，是中国规划展示行业最早的管理成员之一。开创了中国规划展示场馆的市场运作体系和经营管理、讲解接待管理模式。先后从事规划展示场馆的市场推广策划、艺术展览策划、规划专业展览策划、科普教育、公众参与和品牌活动策划。从2001年至今先后参与、主持了200余个国际国内绘画、雕塑、设计等艺术类临展和建筑、规划、城市主题类临展的策划工作，并承担了2011年以来上海城市规划展示馆规划专业展示的主创和文案编纂工作。

文

翁文斌+佟鑫[城市中国]+刘雅妮[城市中国]

图片提供

上海城市规划展示馆

漫步街巷 感受城市温度
风貌保护道路影片

Basic Information

Theme: "Iconic Shanghai" Series Exhibition

Time: 9 September 2017 – 28 January 2018

Venue: Shanghai Urban Planning Exhibition Center (No.100 People Avenue, Huangpu District)

Host: Shanghai Urban Planning Exhibition Center

Organizers: Tongji University College of Architecture and Urban Planning, Shanghai Surveying and Mapping Institute, Shanghai Youth Activity Center, Shanghai Shoubai Cultural and Art Co., Ltd., Shanghai Anmoji Architectural Planning and Design Co., Ltd., Shanghai Jiaotong University Urban Renovation, Conservation, Innovation Research Center, Nanjing University Press, Shanghai Jinshan Peasant Painting Institute, Shanghai Dibby Cultural Development Co., Ltd., etc.

Curators

WENG Wenbin

Chief engineer of Shanghai Urban Planning Exhibition Center(SUPEC). Having joint SUPEC since January 2000, he is one of the earliest members of China's planning exhibition industry, who has created the market operation and management system of China's planning exhibition venues, and introduced the model of guided tour. He has undertaken a variety of roles in marketing, art curation, urban planning exhibition curation, public education, public events, and branding events. Since 2001, Weng has participated in the curative work of more than 200 temporary exhibitions worldwide, covering painting, sculpture, design, architecture, urban planning and urban-related themes. Starting from 2011, he has been a curator, contributor and editor for urban planning projects in SUPEC.

Text

TONG Xin [Urban China] + LIU Yani [Urban China]

Image Source

Shanghai Urban Planning Exhibition Center

《舟山路旧时光》油画　尺幅：65cm╳91cm　作者：孙家珮　馆藏作品

《武康大楼》 作者：曹恒律

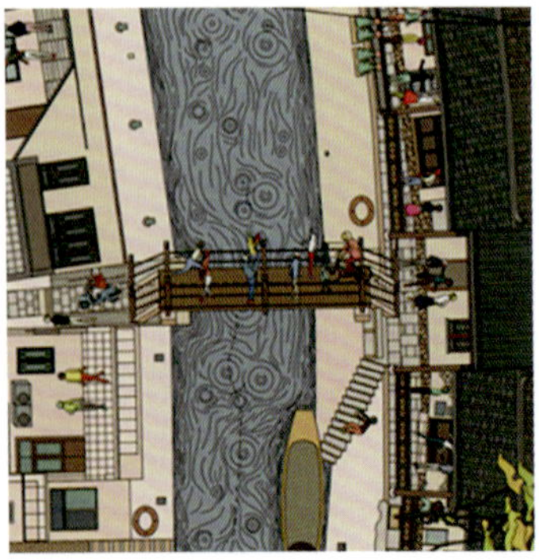

《新场古镇》全景轴测图 创作：绘造社 展览现场以16m × 2.2m的新媒体动画绘本形式呈现了全长1500m的新场大街为核心的上海新场古镇历史文化风貌保护区动态全景图，堪称现代版的“清明上河图”。该展项由绘造社与上海安墨吉建筑规划设计有限公司共同发起创作。

Historic Town of Xinchang: A Panorama *created by Drawing Architecture Studio.*

The work presents the entire protected historic town of Xinchang, with the 1,500m eponymous street as its axis, using the novel media of a 16m × 2.2m animated scroll echoing the famous Qingming Shanghetu (Riverside Scene at Tomb-sweeping Festival). *It is co-created by Drawing Architecture Studio and AMJ.*

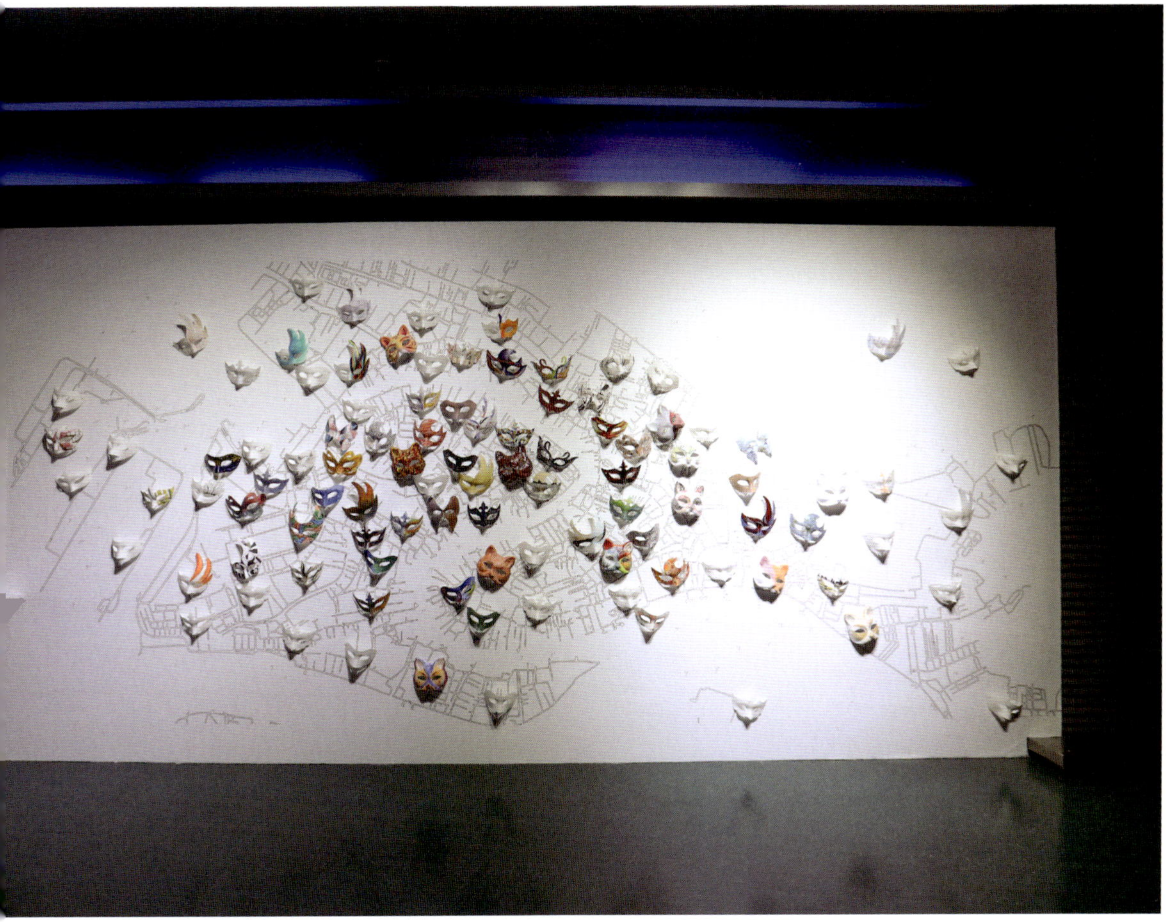

《连接—城市公共空间的多元视角》展览照片

为风貌而设计：徐汇区岳阳路实践案例展

DESIGN FOR THE HISTORIC AND CULTURAL SCENE: SITE PROJECT OF YUEYANG ROAD, XUHUI

策展主题：为风貌而设计

EXHIBITION THEME: DESIGN FOR THE HISTORIC AND CULTURAL SCENE

项目简介

徐汇区天平街道的岳阳路，位于衡山路—复兴路历史文化风貌区南端，为一类风貌保护道路。天平路街道位于徐汇区东北部，东起陕西南路，西抵天平路、华山路，南临肇嘉浜路，北依淮海中路、复兴中路。街道面积2.68平方千米，含多条一类风貌保护道路、多个文物保护单位及优秀历史建筑，有机融合了近代法租界西区风貌。在城市有机更新中，徐汇区天平街道力图实现建筑可以阅读，街区适合漫步，城市始终有温度，更好地展现天平街道所代表的徐汇风貌的魅力。

策展主题阐释

2017年“为风貌而设计：岳阳路实践案例展”紧紧围绕上海作为国家历史文化名城的使命，加强城市历史文化遗产和非物质文化遗产保护和传承，积极推动城市更新。“历史、风貌、文化、艺术”是本次展览的主题策展理念。岳阳路的公共空间是历史人文的聚焦点和缩影，是大众认知历史风貌的重要空间场所。其新旧交融、有机拓展、传统与当代融合的表象，都直接反映了历史、风貌变迁与传统延续的矛盾、冲突与交融，独具魅力与当代学术意义。空间艺术季强调公共空间发展的时间线和历史人文的发展面，使它们直面并相融，同时体现了当代人文、生活、风貌的典型特质与衍变过程。因此，策展的重点是寻找“历史风貌提升”与“文化兴市”有机融合的切入点，创造城市景观与历史文化交融的机会。

展示内容

“岳阳路实践案例展”由风貌影像展、历史文化展、天平名人展、互动设计音乐展等内容组成，通过公共空间的艺术语言，传递人与物之间更直接的关系。其中摄影展展示了19位艺术家和1个创作团队为上海风貌区量身定做的高水平作品100件，一半以上的作品是邀请艺术家直接驻现场进行创作，参与的艺术家包括顾铮、罗永进、杨振中、杨福东等；历史文化展以天平街道为核心区域，主要展览了亚历山大•赉安（Alexandre

Leonard）创作的建筑作品；天平名人展则以文化沙龙的形式呈现历史上本地区的名人及其生活场景，并纳入当代社区活动；互动设计音乐展主要是为期5天的现场音乐演出，邀请本土资深吉他手汪文伟、中国美术学院上海设计学院副教授於阗及其他前沿的创新音乐人。通过不同展区及其展示内容，将风貌街区环境整治与景观品质提升的成果带入市民的现实生活。

PROJECT INTRODUCTION

Yueyang Road in Tianping Sub-district, Xuhui District is a first-class preservation road situated at the south end of Hengshan Road-Fuxing Road Historical and Cultural Landscape. Tianping Sub-district Community lies in the northeast of Xuhui District, bordering South Shaanxi Road (East), Tianping Sub-district and Huashan Road (West), Zhaojiabang Road (South), Middle Huaihai Road and Middle Fuxing Road (North). The area of 2.68 square kilometers contains multiple first-class preservation roads, protected historical and cultural sites, and outstanding historical buildings, making it one of the most representative areas in west sector of Shanghai French Concession. The cultural and historical landscapes are organically integrated with the blocks thanks to the methodology of "micro regeneration, micro improvement and micro design" and the long-term, stage-wise implementation. In the organic regeneration process, the project aims to create "readable" buildings, pedestrian-friendly blocks, and make Tianping Community a warm place that fully demonstrates the enchantment of Xuhui District.

THEME INTERPRETATION

This site project highlighted efforts to protect, utilize and regenerate historical sectors and intangible heritage of Shanghai, a city known for its past. It centered on several keywords, namely history, landscape, culture and arts. As an important cross point for people to perceive and understand the culture in the context of history, Yueyang Road public space is the epitome of public life. The integration of old and new, flexible expansion, as well as the representation of tradition and contemporary in Yueyang Road directly reflect the conflicts embedded in the transition from past to the present, the continuation of tradition with the transformation of the contemporary. SUSAS emphasized on the aspects of time and humanistic concern during the progress of public space, creating a platform for dialogue and further collaboration, at the same time showcasing the transition process of contemporary life and culture. It was the focus of this project to find a way to properly integrate the enhancement of historical landscape and regeneration of city culture, to create opportunities where city landscape and historical culture meet and communicate.

EXHIBITION

The site project included four sections: Landscape Photography Exhibition, Culture & History Exhibition, Celebrities of Tianping Sub-district, and Interactive Music Show. In the photography section, 100 photos were displayed, many of which were created specifically for the site. Invited artists included GU Zheng, LUO Yongjin, YANG Zhenzhong, YANG Fudong and others. The History & Culture section focused on buildings in the sub-district, in particular those designed by Alexandre Leonard. Celebrities of Tianping Sub-district showcased local celebrities and their life scenes in the form of culture salons. The five-day Interactive Music Show invited a number of local musicians, such as local guitarist WANG Wenwei, and YU Tian, vice professor of China Academy of Art Shanghai Institute of Design. These exhibitions were designed to showcase the achievements of community and landscape improvement to a wider range of the public.

传统的建筑展览，形式无外乎展板、模型或者视频，妥善安装后邀请观众观看。但是，这些做法缺乏交互性，内容也往往显得过于生涩和呆滞。观众，特别是一般观众，似懂非懂的展览体验，在自己与建筑专业之间立起了高墙。近年来，创新型展示方法兴起，如营造公共事件使观众卷入其中并产生共鸣，让他们在潜移默化中了解专业设计所要传达的信息。2017上海城市空间艺术季可谓是这种趋势的探索实践之一。主展场由旧有的大型工业遗存——筒仓改造而来，除了作为承担各种展览活动和公众讲座的空间载体，建筑本身就是景观。如此将建筑实践作为展品的做法，在建筑展览当中日趋普遍，徐汇区岳阳路实践案例展的做法与其形成一定呼应。

“为风貌而设计：徐汇区岳阳路实践案例展”前后历时两月有余，共五场活动，包括四场展览和一个音乐周，其中三场展览作为案例展主体贯穿展期始终。展览的三处场地坐落于徐汇区衡复历史文化风貌保护区，包括复兴中路1331号黑石公寓、永嘉路578号永嘉新村正门口的偏房和永嘉路511号原徐汇区天平街道办公室。经策展团队改造，它们成为城市微更新和公共艺术之间的衔接实体，引发了一场关于如何在公共空间中合理进行艺术展示、组织建筑展览的思考与实践。

Traditional architectural exhibitions with the form of panels, models or videos are properly installed to invite viewers to watch. However, these practices lack interactivity and the contents tend to be too suffocating and sluggish. The audience, especially the general public, may not fully understand the contents of exhibition so that a high wall is set between themselves and the exhibitions. In recent years, innovative display methods have emerged, such as creating public events to engage and resonate with the audience, allowing them to understand the information that professional design conveys in a subtle way. The 2017 Shanghai Urban Space Art Season is one of the explorations of this trend. The main exhibition site was transformed from the silo, which was a former large-scale industrial relic. In addition to being a space carrier for various exhibitions and public lectures, the building itself is a landscape. The form of using architectural practice as an exhibit is becoming more and more common in architectural exhibitions. The case exhibition of Yueyang Road in Xuhui District is in line with it.

The case exhibition of Yueyang Road lasted for more than two months, including 4 exhibitions and a music week in total, and 3 of the exhibitions went as the main body throughout the entire exhibition period. Three venues were chosen for the exhibition and they were all located in the Hengshan Road-Fuxing Road Historical and Cultural Area of Xuhui District, specifically the Blackstone Apartment, No. 1331 Fuxing Middle Road, the house at the entrance of Yongjia New Village, No. 578 Yongjia Road, and the previous Tianping Sub-district Office, No. 511 Yongjia Road. After being transformed by the curatorial team, these three buildings became the interface between urban micro-regeneration and public arts, and triggered a discussion on how to properly display art and how to organize architectural exhibitions in public spaces.

采访：策展人薛鸣华&王林

艺术，公共空间的另一解读视角

这场实践中，多个分展场如何做到统一协调并仍保有各自的特色？

这次我们不仅把艺术作品放置在公共空间中，还进一步思考什么样的艺术形式适合这些公共空间，要在衡复历史风貌区揭示公共空间的特质。我们希望尽量多地用实践体现空间与艺术的关联。艺术是可以解读空间的。人们看到一座美的建筑，会对设计者产生好奇，想知道设计背景、想法、设计者的生平等等。我们决定让影像艺术家用影像作品做一个展览，用照片解读环境、空间和历史，这些历史信息会让人感受到上海丰富的文化积淀。我在展览序言中写道："请艺术家用艺术的眼光看我们平时忽略的、习以为常的空间，告诉我们这些空间的独特之处。"这就是黑石影像展的由来。我们当时考虑在徐汇找一栋典型建筑物办展，委托衡复集团筛选，最后选定了黑石公寓。很多市民都听说过黑石公寓，但从没去看过，很神秘。

黑石公寓建于1926年，是一幢英式钢筋混凝土公寓大楼，现为上海市第四批优秀历史建筑。其因填充墙体和部分构件采用黑色石材而得名。公寓的主立面左右对称，正面底层有一个用科林斯双石柱支撑的较大的挑出门廊。整个建筑多处有弧线装饰，增添了巴洛克风格的韵味。

在了解上海历史风貌时，我们遇到了几个代表性人物，比如建筑师邬达克，还有比较出名的鲍立克。衡复地区的建筑有丰富的历史，法国人当时参与了城市的建造，我们觉得也应该选择一位有代表性的建筑师来解读。经法国驻上海总领事馆推荐，我们与一位研究法国建筑师赉安的民间人士结识。上海有一大批对城市、建筑有热情的民间人士，虽然是业余做研究，但细致度和表达力都很优秀。经过交流和思考，我们放弃了做建筑师民间传记的想法，选择从专业的视角做研究，就从赉安在徐汇的建筑作品入手。用了半年时间，我们拍摄这些建筑，搜集资料，王林教授的团队与城建档案馆合作逐一论证每栋建筑的设计资料。赉安作品的资料档案保存得非常完整，这从侧面说明，上海早就是一个现代化的城市，公共资料保存得非常好。最后我们确认了三十栋建筑，形成相对完整的图文信息呈现。

赉安的这个展览项目体现了上海作为国际大都市的典型特质，于是我们为展览取名“来自法国的上海人：赉安洋行建筑在徐汇”。在做永嘉路景观提升项目时，我们一直跟天平街道共同探讨做城市微更新时打造公共空间的方式，除了景观美化以外能否在功能上做一些引导。上海应该有很多沉浸在生活中的文化空间，我认为这才是公共空间文化意义和艺术内涵的最佳呈现。永嘉路578号永嘉新村门口的房子，此前是闲置空间，微更新过程中我们提出街区应该有一个能让市民交流的空间。天平街道也在想这个问题，他们有很多文化名人资源，是一个很有文化特色的街道。我们专门为了这次展览和后续的社区文化功能，对这个空间进行定向的改造。这样这次展览的另一个层次就展现出来了，用微更新结合展览内容的方式去改造一个空间，这个完整的过程才是我们想要的“案例展”。这是一个相互成就的过程，建筑改造与赉安洋行建筑展的研究同步进行。展览有三个层面：城市微更新，艺术和文化功能注入到社区肌理，文化活动激发市民生活的活力，符合卓越的全球城市应具备的文化丰富性。如果这样的案例能够普及，各个社区的个性就会得到充分呈现，文化的丰富性是上海城市文化的根基所在。

在这个过程中我们看到，艺术是需要空间承载的，除了展示还应该有交流空间。天平街道一直在做文化名人交互活动，他们想到可以腾出一个办公空间，结合景观风貌微更新项目进行改造。于是我们对永嘉路511号进行了内部改造和外部修缮提升，使其成为社区一个重要的文化节点，为居民提供举行戏曲沙龙的场所。这个项目还有一个特别的尝试：二楼还有居民在住，但我们同样对一楼进行了改造。这也启示我们，改造不一定要整栋，分层改造也是可行的。“让文化和艺术走家串户”，为什么不可以楼上居住、楼下做文化活动呢？我们这次改造的两个空间分别有展示和承载两种功能，这跟以往直白地美化一个空间的实践不同，对我们来说也是一个惊喜。

空间新想象的创造——跨界与科技

让小众音乐融入历史建筑的想法基于怎样的考虑？

除了在黑石公寓做黑石影像展以展示徐汇风貌外，我们也想用其他的艺术形式阐释这个空间，把人们吸引过来，让人们了解这个独特的城市空间。附近整个片区包括上海音乐学院、上海音乐学院附中、上海交响乐团等音乐文化机构，是一个音乐聚集区。加之徐汇区需要打造一个以音乐元素为亮点的区域，于是我们尝试在听觉道路上探索。确定音乐、当代性和公共空间这三个要点后，我们和中国美术学院上海设计学院影像与传媒艺术专业研究城市空间与声音的於阆副教授联手。他把声音的元素技术化为影像，使声音与空间产生关系，对声音的演绎也在空间当中展现，从而探索这个空间是否适合这种声音展示。

在公共空间中展示声音有多种形式，而我们的空间不太适合做音乐节，经过共同探讨，最终确定了“声·相”城市风貌音乐周系列表演，包含“弄·堂”“适·意”“白·相”“腔·调”和“搭·界”五个主题，分别在五个晚上演出。最后一场“搭·界”的现场，我们本担心会影响到周边居民，但最终来参加的人都很喜欢，有些人连听了五天，居民反响也比较好。我们觉得这次的尝试非常值得。

以黑石影像展中的VR展示项目为例，VR技术对公共空间的发展会有怎样的帮助？

未来的城市空间和公共空间，不仅存在于人们日常感官触及的地方，也会通过数字技术呈现，AI技术、VR技术是打开未来空间的钥匙。其实微信也是一种空间，所以很难预测未来公共空间的定义会发生怎样的演变。我们尝试研究了虚拟空间，用VR技术对岳阳路一平方千米范围内的空间做了数字化处理。虽然目前的技术只能让一人戴上设备体验，但我想不久就可以实现多人同时体验。

城市更新，从术业专攻到跨学科实践

本次案例展展现出明显的在地性，使历史风貌保护区与社区形成共同体，提出“社区规划师”的概念，它的重要性是什么？

城市的发展有自己的规律，需要有预见性地去探索。一个城市按照正常发展规律，会经历人口的增长或减少，区域的繁荣或衰退。这一过程中，我们能强烈感受到三个核心元素：一是科技改变生活，二是注重文化传承使生活更丰富，三是人们对历史传统更加重视。在专业研究的基础上，我们能了解到城市发展有自己的轨迹，就有办法解决很多问题。比如城市风貌保护，所谓“永不拓宽”就是对历史的尊重，很多风貌保持千年的欧洲城市做到了这一点。因而我们可以预见，未来对风貌保护历史价值的认定应该是这个方向。

短期来讲，社区规划师作为一种新的机制和实践很有意义。城市微更新的实现，需要有专业人士在社区中长期有效地参与和监督。或许不只是社区规划师，未来还会有更多全新的概念出现，比如上海河道管理和河道设计采用的“河长制”。未来各个专业都可能会进一步介入社区，比如社区也会有艺术总监、文化总监。有理由相信，专业化的顾问机制会对社区管理的提升起到重要作用。

建筑事务所、各级政府、学界、艺术界和公众之间的互动合作促成了此次实践案例展，建筑师在其中扮演怎样的角色？其与传统项目的区别是什么？

作为建筑师、设计师、规划师，我们是城市建设的参与者。为谁做，为什么要做，对这些问题的认识在实践中不断深入。城市的建设时期不同，大家注重的点也不尽相同。早期的建筑师是在新城造房子，规划师则主要规划新区、产业区。到了新时期，我们会发觉，城市发展要回归到自身，有很多传统的问题需要解决，还有很多规划、建筑层面的事情要做，提升城市本质才是重要的任务。城市微更新是对早期开发大拆大建过程中出现的问题进行重新审视，是一个客观举动。在新的阶段，我们需要精细化地发掘建成环境以外的内容，以更好的理念作为依据，并丰富我们对工作的理解。从拿到任务书到与甲方探讨任务书，建筑师逐渐深入项目的内涵和实质，以实现真正意义上的完美设计。为了使城市的发展更为完善，我们的参与度与以往相比发生了很大变化。

做了一些尝试后，我们也学会用其他专业的角度，如艺术角度，关注自身专业的局限。一些建筑师已经开始参与前期策划和后期运营的工作，虽然规模不大，项目也不一定完整，但尝试向产业链的前端和后端发展的动作一直在进行。这次城市空间艺术季中有大量的建筑作品，它们只是作为一个载体，而我们可以思考为什么这些建筑会出现，它们的出现会带来什么，这些做法会为建筑这个专业带来什么样的可能性。

我不认为将来建筑、规划、景观行业会出现行业性质的转变，它们依然互相独立，其科学性无法替代。但实践和学科研究有必要做更多跨界理解，不同领域的视角能提高我们工作的高度和综合能力。这次我们也给合作的艺术家们提供了很多启示，他们也从整体的角度去思考艺术的呈现，比如公共性、公众参与、高端艺术如何在民间普及，等等。不管具体做什么，我们还是要互相激发，最终回报这个城市。

人们习惯对公共空间进行归类，但它远比我们想象的要丰富。不同专业引领独特的丰富性是城市文化生命力的典型表现。有些艺术形式走到最后，一定要借鉴其他学科才能走出新的道路。对音乐来说，科技对硬件的提升有颠覆性的作用，音乐的革命一直在进行当中。也许有一天，机器人也会对建筑师的工作形成威胁。但有人说AI介入建筑业，建筑师也会自主革命，用其他形式来自我革新。从其他行业了解自己的弱点，找到完善自己的道路，这是很了不起的事。

INTERVIEW WITH CURATOR XUE MINGHUA & WANG LIN

Art, Another Perspective to Understand Public Spaces

How do multiple exhibition venues achieve uniform coordination and still maintain their respective characteristics?

In addition to considering how to place art works in public spaces, we also thought about what kind of art form was more appropriate to be presented in these public spaces and be able to enhance the striking features of public spaces in the Hengshan Road-Fuxing Road Historical and Cultural Area. We wanted to use as much practice as possible to reflect the connection between space and art. To some extent, art can be used to understand spaces. When people see a beautiful building, they would probably get curious about the architect and want to know more about design ideas and backgrounds. The historical information will trigger people's feelings for the rich cultural heritage of Shanghai. So we decided to invite artists to interpret the environment, space and history by means of videos and photos. As I wrote in the preface for the exhibition: "Look at spaces we usually ignored from an artistic perspective, show the distinctiveness of them." This is the origin of the Blackstone Landscape Photography Exhibition. At that time, we were looking for a typical building to organize the exhibition in Xuhui District, eventually Hengfu Group helped us choose Blackstone Apartment. This building was originally built in 1926 and remains a mysterious place for many people.

Built in 1926, the Blackstone Apartment is a reinforced concrete apartment building of British architectural style, now listed one of the fourth batch of outstanding historic buildings in Shanghai. It was named after the black stone-material used for its infill walls and some of the components. The main façade of the apartment is symmetrical, with a wide porch supported by Corinthian double pillars on the ground floor. There are curved decorations throughout the building, adding a sense of baroque aesthetics.

We noticed some representative figures such as architect Laszlo Hudec and Richard Paulick in our researches. The French were involved in the construction of the Hengshan Road-Fuxing Road Historical and Cultural Area, which enriched the architecture of the area. For this reason, we thought it would be better to choose a representative architect to interpret local buildings. Recommended by French Consulate, we came to know a researcher who's dedicated to studying Alexandre Leonard. There is a group of amateur researchers who are enthusiastic about the city and architecture in Shanghai. They are quite excellent in meticulousness and expressiveness. After a discussion with this researcher, we gave up the idea of doing a biography, and decided to do research from a professional perspective, specifically, starting with the architectural works of Alexandre Leonard in Xuhui District. We spent the next six months on collecting information about his buildings and taking photos for them. Professor WANG Lin and her team worked with the Shanghai Urban Construction Archive to guarantee the accuracy of the materials. The data file of Alexandre Leonard's works was complete and well preserved, which indicated the level of modernization of Shanghai at that time. Finally, we chose thirty buildings and formed a relatively complete graphic information presentation.

This exhibition of Alexandre Leonard's works was titled as The Shanghainese from France - Alexandre Leonard in Xuhui District. It proved Shanghai as an international metropolis. During the process of Yongjia Road project, we had been thinking about the possibility of providing guidance on functions for an urban regeneration project, in addition to creating new public spaces. We believe cultural spaces scattered around and immersed people's daily life are the best presentation of cultural significance and artistic implication of public spaces. There was a vacant house at the entrance of Yongjia New Village and we were thinking an urban micro-regeneration project to provide a public space for people. This idea was echoed by Tianping Sub-district Office. So eventually we transformed the space for this exhibition and also for future community activities. The renovation and the research of Alexandre Leonard's architectural works went simultaneously. The exhibition had three aspects in the end, the urban micro-regeneration, the injection of artistic and cultural functions into the community texture, and the reawakening of community vitality by cultural activities, which meet the needs of cultural richness for an outstanding global city.

Art needs space to spread its meanings, to display and to communicate. Tianping Sub-district had been doing cultural celebrities' communication activities and they wanted to transform an office as the part of urban micro-regeneration project. We carried out both internal renovation and external repair at No. 511 of Yongjia Road, turning it into an important cultural node in the community and providing residents a place for traditional opera events. It's worth mentioning that we only refurbished the first floor, the second floor is still occupied. This means renovation does not have to be the entire building. These two transformed spaces both have functions of display and capacity, and they are different from the past practice of simple beautification.

New Imagination:

Crossover and Technology

How did you come to the idea of live music performance at the historic buildings?

We wanted to attract people to this unique urban space by multiple means other than merely the photography exhibition. This area is home to several institutes including Shanghai Conservatory of Music, the Middle School affiliated to Shanghai Conservatory of Music and Shanghai Symphony Orchestra. Xuhui District government wanted to focus on this and enhance the musical atmosphere, so we decided to explore possibilities from an auditory perspective. Music, contemporaneity and public space are three mains points. Associate Professor Yu Tian from Department of Image and Media Arts of China Academy of Art helped us with this. He transformed elements of sound into images, creating a relationship between sound and physical space. The interpretation of sound was also displayed there to see whether the space was suitable for such a display.

There are many forms to display sound in public spaces. It is clear that our space was not applicable for music festivals, so we finally decided to organize the Urban Environment Music Week, named Music and Phase, with serial performances. Five live shows with different themes, "Long · Lan", "Suit · Desire", "White · Phase", "Tune · Tone", and "Cross · Over" were performed. Audiences enjoyed the shows. Some people came every night, five nights in a row. We were pleased to see our efforts had been worthwhile.

You have used VR technology in the exhibition. What kind of benefits can new technologies bring to the development of public spaces?

Urban space and public space will not be limited to physical spaces in the future; they can be digital as well. AI and VR technologies are more like keys to open up future spaces. In fact, it is hard for us to predict the evolvement of definition of public spaces. The apps we use everyday, like WeChat, can be regarded as a space to some extent. For this project, we studied the virtual spaces and used VR technology to digitize spaces within one square kilometer radius of Yueyang Road. Although at the moment it only allows one person to wear the device to experience the virtual space, I believe simultaneous multi-person experience can be achieved in a very near future.

Urban Regeneration, from Specialization to Interdisciplinary Practice

This case showed an obvious in-situ feature, forming a unity with local society in a historical cultural area. You even came up with the idea of "community planner". What's so important about it?

The development of a city follows its own rules, with its own ups and downs, growth or declines. There are three core elements during this process: firstly, technology changes life; secondly, cultural inheritance enriches life; thirdly, traditions are gaining more attention. Urban histories based on professional researches can help make the city better. For example, the policy of never widening roads can be regarded as the respect for history, through which many European cities have achieved to preserve their millennium city view. We believe the identification of the protection value of historical landscapes should be on this track.

The introduction of community planners as a new mechanism is very significant, because urban micro-generation requires the effective participation and supervision of professionals in the community for a period of time. There will be more concepts as such in the future, for instance, the River Chief System in Shanghai river management and river design. Professionals of different backgrounds are expected to be further involved in the development of communities. Such a professional advisory mechanism will play an important role to improve community management.

> The interaction and collaboration between architectural studios, government, academia, the art world and the public have led to the success of this project, and what is the role of architects in it? What is the difference from your usual architectural projects?

Architects, designers and planners are all participants of urban construction. We all gain a deeper understanding of questions like for whom and why to do it in the process. The focus of architects changes from time to time. In the early days, architects built new buildings and city planners planned new industrial districts or development areas. Nowadays it is noted that urban development needs to look back and focus on itself, on the essence of a city. There are many conventional problems to be solved and urban planning is therefore immature. Urban micro-regeneration is objective conduct to re-examine problems that occurred during the early phase of large-scale demolition and construction. We need to refine the content beyond the built environment, find better ideas as a basis to enrich our understanding of the plan and design work. Architects get involved from the very beginning of a project and gradually understand more of the site to come up with a design. The extent of participation has changed a lot.

We are also looking at the limitations of our profession from perspectives, and some experiments have been done in this regard. A large number of architectural works were shown in SUSAS 2017, as a carrier or a platform to show why these buildings appear, what they bring, and what kind of possibilities they produce for the profession of architecture.

Disciplines like architecture, planning and landscape will still be independent, but more interdisciplinary practices and researches are needed. The perspective from different fields admittedly helps increase the height and comprehensive ability of our work. We inspired our cooperated artists to present art from a more holistic perspective. No matter what kind of work we do, it's important for us to inspire each other and ultimately contribute to the city.

Public spaces are more diverse than we think despite the inclination of being categorized. The pluralistic culture created by different fields is a proof of urban vitality. Some forms of art need to derive inspirations from other disciplines. For instance, the music industry is still undergoing revolution with technologies aiding to upgrade the equipment. For architecture, some say it's in danger because of AI technology, but it's also true that with the involvement of AI architects will be motivated to self reflect and make changes.

展览基本信息

主题：为风貌而设计

时间：2017年10月21日—12月25日

地址：上海市徐汇区复兴中路1331号黑石公寓

上海市徐汇区永嘉路578号

上海市徐汇区永嘉路511号

主办单位：徐汇区人民政府

承办单位：徐汇区规划和土地管理局

协办单位：上海交通大学城市更新保护创新研究中心、上海安墨吉建筑规划设计有限公司

支持单位：上海衡复投资发展有限公司

策展人介绍

薛鸣华

同济大学建筑城规学院景观园林硕士，2015“互动水乡 朱家角 · 尚都里实践案例展”分展区策展人，上海安墨吉建筑规划设计有限公司设计总监。

王林

中国城市规划学会城市更新学术委员会副主任，上海交通大学建筑系教授、博导，中国城市治理研究院研究员、城市更新 · 保护 · 创新国际研究中心主任，2015“更 · 新场——浦东新场古镇实践案例展”分展区主策展人。

文

佟鑫[城市中国]+王卉[城市中国]

图片提供

上海安墨吉建筑规划设计有限公司

Basic Information

Theme: Design for the Historic and Cultural Scene

Time: 21 October 2017– January 2018

Address: Blackstone Apartments (No. 1331 Middle Fuxing Road);No. 578 and 511 Yongjia Road,Yongjia New Village

Host: People's Government of Xuhui District, Shanghai

Organizer: Xuhui District Planning and Land Authority

Co-organizers: Shanghai Jiao Tong University City Lab Shanghai AMJ Architecture & Urban Planning Design Co., Ltd

Supporter: Shanghai Hengfu Investment & Development Co., Ltd

Curators

XUE Minghua

Postgraduate of Landscape Architecture of the College of Architecture and Urban Planning Tongji University, curator of 2015 Interactive Waterland Zhujiajiao-Shangduli section, Director of Design of Shanghai AMJ Architecture & Urban Planning Design Co., Ltd

WANG Lin

Deputy director of Regeneration Academic Committee Urban Planning Society of China, professor and Ph.D. supervisor of Architecture Department, Shanghai Jiao Tong University, researcher of China Institute of Urban Governance, director of International Research Center for Urban Regeneration and Protection, chief curator of 2015 Regeneration Exhibition Area in Xinchang Ancient Town Pudong New Area Shanghai.

Text

TONG Xin [Urban China] + WANG Hui [Urban China]

Image Source

AMJ

黑石影像展，有故事的展品与有历史的展厅相辅相成
Blackstone Landscape Photography Exhibition was a combination of artwork and exhibition hall

折衷主义风格建筑，黑石公寓
Blackstone Apartment, an example of eclectic architecture

黑石影像展，有故事的展品与有历史的展厅相辅相成
Blackstone Landscape Photography Exhibition was a combination of artwork and exhibition hall

永嘉路588号，赉安旧居
No.588 Yongjia Road, former residence of Alexandre Leonard

复兴西路34号，卫乐公寓
Weile Residence at No.34 Fuxing West Road

高安路72号，住宅
A residencial building at No.72 Gaoan Road

建国西路394号，高层公寓住宅
High-rise apartment at No.394 Jianguo West Road

华亭路71弄1-7号，住宅
A residecial building at No.1-7, Lane 71, Huating Road

复兴西路26号，住宅
A residencial building at No.26 Fuxing West Road

从赉安及赉安洋行的建筑设计中，感受上海近现代新风格建筑
A new architectural style in modern Shanghai can be noticed in Alexandre Leonard' s works

"白 · 相" "White · Phase" Show

"搭 · 界" "Cross · Over" Show

"弄 · 堂" "Long · Lane" Show

"腔 · 调" "Tune · Tone" Show

以"声 · 相"造"境"、"还原风貌"，扩展空间理念，让观者进入此"境"，引发共鸣
"Music · Phase" Show created a scene and an atmosphere for the audience to dive in and become emotionally involved

双重曝光·旧厂新生·静安区新业坊实践案例展

DOUBLE EXPOSURE, FACTORY REBIRTH: SITE PROJECT OF INNOVATION GALAXY, JING'AN

EXHIBITION THEME: DOUBLE EXPOSURE, FACTORY REBIRTH

策展主题：双重曝光·旧厂新生

项目简介

静安新业坊的前身为创建于1959年的上海冶金矿山机械厂，是彭浦工业区乃至闸北地区工业发展的见证，是上海近代工业向现代工业迈进的历史缩影。独具韵律的大跨度建筑作为工业遗产兼具丰富的文化和历史价值。未来静安新业坊以工业历史文脉的有效保护和再利用为目的，以打造未来的文化地标为目标，建立从演播室、制片人、导演工作室，到IP孵化、投资交易、衍生品开发等覆盖一系列内容的创作产业链。静安新业坊以创新为重点，同时保留历史的痕迹，鼓励公众参与，将生产与生活紧密地融合在一起。

策展主题阐释

“双重曝光”是摄影术语，原意指拍摄时在同一张底片上重复曝光两次，让各自独立的影像重叠显示在同一画面上。从生产机械设备、发展城市硬实力的工厂，到创造精神文化价值、为公众提供软服务的“工厂”，新业坊即将完成其作为城市公共空间的历史转身。静安区新业坊实践案例展摒弃了常规的展览模式，于2017年10月、2018年1月在上海冶金矿山机械厂的废旧空间和改造后的新业坊空间进行了两次艺术快闪活动，向公众展示跨越时空维度的价值重合。

展示内容

“静安区新业坊实践案例展”围绕“双重曝光·旧厂新生”的主题，分别通过两期“咱们工人有力量”和“咱们人人有力量”，配合艺术装置、艺术市集和音乐派对等具有高度参与性、互动性的艺术活动，使观展的市民切身感受到在地化创作和公共艺术实践给城市空间带来的提升及改变。

2017年10月21日举办的艺术活动以“咱们工人有力量”为分主题，由艺术装置、艺术市集和露天电影三部分组成。艺术家以厂区里废旧的材料为灵感，创作地景艺术、霓虹装置、声音装置等形式丰富的新媒体艺术，为公众营造浸入式工业体验，实现公众与空间和历史的连接。艺术市集抛开传统的售卖形式，注重艺术体验和市民的参与性。以复古混搭、现代创意为主题元素，包含工人食堂美食、艺术体验工作坊、传统手工艺展示等各种活动。夜晚还有露天电影的怀旧体验，让观众在参与互动的同时，完成不同时代的人的连接。2018年1月6日的活动以“咱们人人有力量”为分主题，更偏向新空间整体设计感和未来感的营造。从充满实验性的学生剧目表演，强调参与性和互动性的街头表演，到全民参与的独特工厂音乐节，都帮助公众塑造了新业坊转型为影视文化产业基地的新认知。

PROJECT INTRODUCTION

Jing'an Innovation Galaxy can be traced back to Shanghai Metallurgical Machinery Factory founded in 1959, which witnessed and represented the industrial progress of Pengpu, and more broadly, the entire Zhabei area. The uniquely large span architecture embodies great cultural and historical values, based on which the new Jing'an Innovation Galaxy intends as an innovation engine of Shanghai while keeping the old industrial flavor intact. As a future cultural landmark, the Galaxy houses various creative industries from theatres, studios to IP incubation and investment transactions. Focusing on innovation while retaining historical traces, Jing'an Innovation Galaxy will motivate public engagement and the integration of economic activities and lifestyle.

THEME INTERPRETATION

Double exposure is a photography jargon, which means that superimposing two images on one film by exposing it twice. A profound and historical transformation is underway of a factory which houses manufacturing machinery and strengthens the city's tangible forces to Jing'an Innovation Galaxy which provides intangible, that is, spiritual and cultural, values and services to the public. Breaking through the old-fashioned exhibition model as guided by educated professionals, two flash activities were held at the old Shanghai Metallurgical Mining Machinery Factory and the renovated Jing'an Innovation Galaxy in October and December 2017 with a common theme "Double Exposure, Factory Rebirth". They showed the public how values miles and decades apart may co-exist in a temporal-spatial unity.

EXHIBITION

Focusing on the theme "Double Exposure, Factory Rebirth", Jing'an Innovation Galaxy launched two sessions of Art Exhibitions, that is, "We Workers Have Power" (October 2017) and "We All Have The Power" (January 2018), together with participatory and interactive art installations, art markets and music parties. These localized public events were designed to inspire visitors with the concept that art transforms and sublimates our beloved city.
The activities in October combined art installations, art market and outdoor movies in a retrospective sub-theme "We Workers Have Power". Inspired by industrial waste, the artists had created various forms of art including land art, neon sign installation and sound installation. Such immersed industrial experience connected people with space and history. The 50-vendor art market featured experience and participation rather than buying and selling. Combining vintage and modernism, the market introduced to its visitors a wide range of activities, such as worker's canteen, art workshops, traditional handicraft and outdoor movies, compacting generations of memories into a single place. On 6 January, empowered by new functions of the space, another sub-theme "We All Have The Power" took over, offering events with elements of film culture and performing arts as the connective medium, so as to motivate public involvement. It also reshaped the perception of Jing'an Innovation Galaxy as a film/cultural base through experimental drama, interactive street shows and unique "crowd-involved" music festival in the renewed "factory".

十年前，静安新业坊是上海原闸北地区传统的大型机械制造基地——上海冶金矿山机械厂所在地。上海冶金矿山机械厂创建于1959年，由多家企业合并而成，曾是上海原闸北区重要的发展动脉。冶金厂从建厂开始就取得了多项技术突破，其试验成功的多项设备在全国居于领先地位，其生产的多项设备填补了国内该领域的空白。鼎盛时期，其所属的彭浦机电工业区产值曾占到全上海产值的四分之一。

2009年，汶水路210号厂房的工业生命在产业改革中结束。同时，伴随闸北、静安两区的合并，这片居大宁商圈以北、紧邻中环线和南北高架的厂房成为城市格局更新的重要节点。一方面，厂区内两栋大体量厂房被列入上海新工业遗产名录，厂区从建筑风貌、环境、景观等物质层面到历史、人文、工业文明等非物质层面都具有丰富的遗产价值。另一方面，随着大宁商圈成为上海市第十四个市级商业中心建设的推进，以及环上大国际影视产业集聚区的发展概念逐渐成形，处于核心位置的厂区亟须凸显区域价值，帮助北上海区域重新导入人气，树立形象。在此背景下，静安新业坊诞生，由华东建筑设计研究总院进行改造，临港集团跟进运营。通过对旧工业建筑和空间景观的改造优化，以及文化创意功能和还建/临建新单体的植入，这里被规划成为历史与未来共生的都市创意文化园区。上海于2010年成为中国继深圳后第二个加入联合国教科文组织“创意城市网络”的城市，并被授予“设计之都”的称号，上海创意产业正处于迅猛发展的新阶段，静安新业坊的出现，正是创新与探索并行的结果。

文化创意园区的迭代发展

1999年，台湾设计师登琨艳将上海苏州河南岸一处旧仓库改建成设计工作室，初期的文化产业形态就此萌发，“旧厂新生”的改造理念随之被广泛使用。根据上海市文化创意产业推进领导小组的数据，截至2017年，上海各式各样经审批的创意园区已达128家。其中，包括同乐坊在内的多家园区早在2005年就采用了类似的改造形式。在相关政策的鼎力支持下，产业园区及其带动的创意产业在上海的发展有目共睹。据上海文化创意产业发展报告，2006年，上海创意产业总产值达2291.71亿元，占全市GDP比重6.55%。十年后，这两个数据分别为9451.6亿元和12.1%。2015年，上海文化产业发展指数也首次超越北京，成为全国第一。

然而，经过近20年的扩张，创意园区数量和质量不匹配的矛盾也日渐凸显，很多园区因经营不善而萎靡甚至关闭，园区兴建风潮有所回落。一方面，一些文化创意园区只提供硬件设施，缺乏配套的服务；另一方面，园区内部管理和运营缺乏整体设计规划，导致大多数园区形态、内容相似性极高。换言之，创意产业园区在看似如火如荼的发展表征下，缺乏自身文化场域的实际内核。

在园区亟须转型的关口，对于包括静安新业坊在内的新晋创意园区而言，如何提升产业认知、加强运营能力、明确园区定位并制定健全的管理政策，将是未来园区发展的必修课。

产城融合，城市更新的一种姿态

静安新业坊位居北中环，是环上大国际影视产业区的一部分，也是北上海差异化产业布局，提升区域价值的关键。不同于上海东、西、南部的发展，中心城区以北的上海在相当长的一段时间中显得有些不温不火。对比虹桥枢纽商圈、迪士尼国际旅游度假区，一直到张江高科技园区和浦东国际机场，甚至松江、九亭等区域板块的不断更新，北上海在城市扩展和产业升级进程中相对滞后。北上海亟须打破这种格局，而人群较为集聚、商业服务较为完备的大宁地区，顺势成为北上海区域发展的先行地。2014年，上海温哥华电影学院在大宁区域成立；2015年，静安、闸北“撤二建一”合并，大宁地区也成为静安文创产业向旧闸北延伸的桥梁。《上海市城市整体规划（2017—2035年）》中明确提出，将环上大影视产业基地作为上海发展文化软实力的重点建设项目。自此，北上海明晰了以文化创意产业带动整个区域转型的发展路径。一方面，弥补北上海在文化创意产业上的不足，且在空间上打破创意产业园区以延安路为轴，聚集在徐汇、虹口、长宁的旧有态势；另一方面，区别于中环线上其他商务区，以影视产业定位，也能使大宁地区所代表的北上海在内中环区找到自己独特的产业立足点。

对于大宁北部的彭浦工业区，静安新业坊的出现是地区借助产城融合而进行城市更新的新标志。具体而言，是以产业规模化发展为契机和保障，通过与周边城区的融合互动，实现产业和城市空间的整合发展。工人新村承载着一代人难忘的历史文化记忆，同时也成为城市更新和社区功能提升的重点对象，如何让两者结合共生，成为城市可持续发展的重点。2000年以后，“旧厂新生”成为上海保护和再利用产业遗存建筑的主要方式，这为文化创意产业的培育提供了先期的硬件资源，也完成了产城融合中城市助推产业的第一步。

据新业坊招商运营副经理陈雪晴介绍，静安新业坊的开发工程分为城市更新与城市开发两部分。前者的重点是选择保留在区域内的、平衡功能和年代等不同特点的老建筑，借此展示上海工业发展过程中的建筑元素，从而沉淀空间记忆。随着城市产业结构的调整，集聚资源的城市核心区成为文化创意产业及以智力为核心的新兴产业的关键发展要素。换言之，从前开设在城市外围的产业园区需要回到城市中心。另一方面，城市在扩张，曾经的城市“外围”实际上逐渐成为城市的新中心，大范围开发的机会所剩不多，以“点”开发取代以“面”开发成为新的趋势。在这个过程中，新业坊的目标，也逐渐从产业发展的推动者转变为城市更新的建设者。

文化产业是城市更新的新动能。通过挖掘在地文化，借助公共空间的形式，实现园区与当地社区更多的融合和互动。正如前文所言，静安新业坊是静安区发展影视资源全产业链中的重要一环，其改造方案也与影视产业息息相关。例如，原有厂房结构被保留，未来建设为SMG的大型演播室；厂区内有上海最长的厂区林荫大道，未来也将成为影视产业展示长廊。加之大宁区域以及万荣路沿街的产业形象已有一定的规模，静安新业坊的开发策略之一，即与所在区政府签署战略开发协议，按照区域本身产业规划方向，发展包括电影、电视和音频产品等在内的视听产业。借助如SMG这样的龙头企业的资金实力，让硬件基础设备在园区内实现共享。“影视制作行业需要重资产投入，影视拍摄设备昂贵且更新换代快。共享设备有助于小型创作公司降低制作成本；对于龙头企业而言，也可以提高资产的利用率。”陈雪晴解释说。

以展激活，创意园区的实践思考

此次上海城市空间艺术季举行的两期（1.0与2.0）静安区新业坊实践案例展，正好是静安新业坊作为创意园区推进融合互动的创新做法。“文化创意产业最终面向C端，即人群本身，所以园区希望在进行产业集聚和内容创作的同时，让内部活动如提供演出内容等可以做到服务周边，保留建筑本身也是为了在文化记忆上吸引群体。”陈雪晴说道。借助如“咱们工人有力量”这样有着鲜明时代特色的口号，2017年10月开展的实践案例1.0首先在叙事上联结公众和老建筑。以其中的“艺术共和”板块为例，通过收集和征集老厂房标语，并将其制作成霓虹灯牌隐藏在展区内，鼓励民众寻找并进行多种形式互动，通过回忆来吸引周围老工人，这种怀旧感的营造，让他们愿意参与互动。在这些氛围的营造和联结构成之际，新型艺术也在展出。

通过使用与工业、工厂相关的主题或材质，贴近记忆的拥有者，让他们改变对场地功能的认知，认同新空间。作品《脉搏》通过使用球体装置，结合老机械布景，来完成对于机器时代的隐喻和对工厂功能的表达。“我们希望公众参与进来，”策展团队代表马慧婷坦言他们的初衷，“从生产设备的工厂到未来的影视文化园区，避免生硬的植入，用更‘润物细无声’的方式，在调性和气质上让大家自己理解和探索空间性质的变化。”

在2018年1月举办的实践案例2.0中，主题进一步拓展为“咱们人人有力量”，试图在人群方面寻求突破。不仅仅是工人，新的个体同样被接纳并被允许以新的方式与空间发生关系。改变1.0的怀旧主题，2.0实践为工业元素注入具有未来感与现代感的设计，如《Zao》系列第二弹，更换第一弹意指“造血、制造”的“造”字为燥起来的“燥”，利用声光电效果立体呈现字意、字体的变形。“在这个年代，人人都是创造者。展览的效果是希望观众不仅仅是观众，而是可以参与其中，甚至成为艺术家。而艺术家也可以作为参与者去体验。”因此，2.0实践的特色圆棚设计效果远超预期，这让策展团队惊喜不已。展览结束后，团队内部总结经验，“在圆棚这样的半封闭空间中，艺术家对其有一定的领地感，和开放式的市集相比，他们会更愿意主动维护领地气氛。对于那些注重参与感，像‘天黑请闭眼’这样的相对先锋的实验表演，封闭式空间也能帮助实现最佳效果，进入棚内的观众受到外界干扰的因素也相对变少。”

诚然，上海不缺少工业遗产转型升级的案例，但是以这样的形式让公众介入改造的实践却不多见。尤其是两个阶段的展览着重强调了新旧转变过程，改造前与改造后带给参观者截然不同的内容记忆点，但因为所处空间相同，记忆也可相互串联。这也正是主题“双重曝光”的意义——借助一系列艺术改造和重置，让民众在跨时空的场景下，获得对厂房工业精神的认同，并将这种认同扩大至更广空间范围。两次案例展参观人数都达到了较预期的五倍之多，无论是静安新业坊团队还是罗浮紫策展团队，双方的积极配合以及一致的理念也为这次实践案例展关于创意园区未来合作发展方法带来更多思考。

产城融合的城市更新需要建构对地区历史文化的认同感，这是产业借以发展的基础元素，也是城市再生的关键所在。对于在开放前以展览的方式，先见性地与公众建立联系的静安新业坊来说，这次展览表明它作为创意园区，期望达成不同于其他园区的更大价值或者更高目标——不仅仅是产业的高质量集聚，更是作为一种空间句法，助力在地城区功能和社区公共服务的提升。如此，静安新业坊能否完成其促进产业发展之外重焕城市活力的任务？展览，仅仅是一个起点。

Ten years ago, the location of Jing'an Innovation Galaxy was the reputable Shanghai Metallurgical Mining Machinery Factory. Founded in 1959 through merger of several enterprises, it was once an important development artery in the former Zhabei District of Shanghai. The metallurgical factory achieved a number of technological breakthroughs in its history, many of its successful tests had taken the lead in the country, and its production had filled the gap in this field in China. In its heyday, the output value of Pengpu Electromechanical Industrial Zone which it belonged to accounted for one quarter of the total output value of Shanghai.

In 2009, the industrial life of No. 210 Wenshui Road ended in the industrial reform. At the same time, with the merger of the two districts of Zhabei and Jing'an, this factory building, located to the north of Daning business district, adjacent to the Central Line and the north-south elevated road, became an important node for urban spacial renewal. On the one hand, two large-scale workshops in the plant area were concluded in Shanghai New Industrial Heritage List. The plant area had rich heritage values from physical aspects such as architectural style, environment and landscape to non-physical aspects such as history, humanities and industrial civilization. On the other hand, as Daning business district became the 14th municipal-level commercial center in Shanghai, and the concept of the international film and television industry cluster around Shanghai University gradually took shape, the factory with its crucial location needed to improve its regional value and help North Shanghai region regain popularity and establish a good image. It was in this context that Jing'an Innovation Galaxy was born, transformed by East China Architectural Design & Research Institute and operated by Lingang Group. Through the transformation of old industrial buildings and landscapes, as well as the injection of cultural and creative functions and the monomers being built or to be built, it was designed to become an urban creative culture park with historical and future symbiosis. In 2010, Shanghai became the second city in China to join the UNESCO "Creative City Network" after Shenzhen, and was awarded the title of "Design Capital". The creative industry in Shanghai is in a new stage of rapid development. The emergence of Jing'an Innovation Galaxy is the result of this innovation and exploration.

ITERATIVE DEVELOPMENT OF CULTURAL AND CREATIVE PARKS

In 1999, Taiwan designer Deng Kunyan renovated an old warehouse on the south bank of Suzhou River into a design studio, sprouting the cultural industry form of Shanghai in the early days. The renovation idea of "Factory Rebirth" has been gradually used since then. According to the statistics from Shanghai Cultural and Creative Industry Organization, by 2017, Shanghai had a total of 128 approved creative parks, and many of them, including Tonglefang, adopted a similar way for renovation in 2005. With the full support of relevant policies, the development of industrial parks and its creative industries in Shanghai is obvious to all. According to the Shanghai Cultural and Creative Industry Report, the total output value of Shanghai creative industries in 2006 reached 229.171 billion yuan, accounting for 6.55% of the city's GDP. And after ten years, these two figures were 945.16 billion and 12.1% of the GDP. In 2015, the Cultural Industries Development Index of Shanghai also surpassed Beijing for the first time surmounting in the country.

However, after nearly 20 years of expansion, the contradiction between the number and quality of creative parks becomes increasingly prominent. Many parks being sluggish or even closed due to poor management has caused the construction of creative parks fall back. On one hand, obvious problems in some cultural and creative parks are the lack of supporting facilities, providing hardware facilities only. On the other hand, the lack of overall design planning for the internal management and operation of the parks resulted in a high degree of similarity in styles of most parks. In other words, creative parks lack the actual core that can build their own cultural field in this seemingly intensive developmental phenomenon.

In face with the urgent need for transformation of industrial parks, for these new creative parks including Jing'an Innovation Galaxy, how to improve industrial awareness, strengthen operational capabilities, clarify the park's position and formulate sound management policies will be a must for the future development of creative parks.

CITY-INDUSTRY INTEGRATION, A MODE OF URBAN REGENERATION

Jing'an Innovation Galaxy, located at the north of Middle Ring Road, is a part of International Film Zone, as well as a key to differentiated industry layout and regional value promotion in North Shanghai. Different from the development of other parts in Shanghai, the area in the north to the city center has developed "tepidly" for a long time. Compared to Hongqiao Central Business District, Shanghai Disney Resort, Zhangjiang High-Tech Park, Pudong International Airport, and even the continuous regeneration of Songjiang and Jiuting area, North Shanghai is relatively behind in the progress of urban expansion and industrial upgrade. In urgent needs to break this pattern, Daning area with relatively more condensed population and more complete commercial services has taken the advantage to become the pioneer of regional development in North Shanghai. In 2014, Shanghai Vancouver Film School was established in Daning area; in 2015, with the merging of Jing'an and Zhabei district, Daning area also became the bridge part of the Jing'an cultural and creative industry extending to the old Zhabei. Shanghai Master Plan (2017-2035) clearly proposed that International Film Zone would be a key construction project for Shanghai's development of cultural soft power. Since then, North Shanghai has clarified the development path of leading transformation of the entire region with cultural and creative industries. It could, on one hand, make up for the shortages of North Shanghai in cultural and creative industries, and break the original spatial pattern that creative parks are clustered in Xuhui, Hongkou and Changning districts with Yan'an Road as the axis. On the other hand, different from other business zones on the Middle Ring Road, the positioning as film & television industry will also enable North Shanghai, represented by Daning area, to find its own unique industrial foothold in the Inner-Middle Ring Road area.

For Pengpu Industrial Zone in the north of Daning, the setup of Jing'an Innovation Galaxy is a new representation of urban regeneration through city-industry integration. To be specific, it takes the scale development of industries as an opportunity and guarantee, and realizes the integrated development of industry and urban space through the integration and interaction with surrounding urban areas. Workers' villages carry the unforgettable historical and cultural memories of a generation, and at the same time become a key target of urban regeneration and community function improvement. How to combine the two together becomes the focus of sustainable urban development. After 2000, "factory rebirth" has been the main approach to protect and reuse the remaining industrial sites in Shanghai, which provides proceeding hardware resources for cultivating cultural and creative industries, and also completes the first step of city's boosting industry during the city-industry integration.

According to Chen Xueqing, deputy manager of the investment and operation of Xinyefang, the development project of Jing'an Innovation Galaxy are divided into two parts, urban regeneration and urban development. The former focuses on the selection of old buildings to preserve, with different characteristics such as functions and ages, in this way showing the architectural elements in the process of Shanghai industrial development and thus precipitating spatial memory. With the adjustment of city industrial structure, city core area of agglomerated resources has become a key development factor for cultural and creative industries and emerging industries with intelligence as the core. In other words, industrial parks that were previously set up on the outskirts of the city need to return to the city center. On the other hand, as the city is sprawling, and the "outskirts" of the city has gradually become a new center of the city. There are not many opportunities for large-scale development and development by "rather than" dimension" has become a new trend. In this process, the aim of Jing'an Innovation Galaxy has transformed to the constructor of urban regeneration from the promoter of industrial development.

Cultural industry is the new kinetic energy in urban regeneration. By exploring the local culture and using the form of public space, more integration and interaction between the parks and the local community can be realized. As mentioned before, Jing'an Innovation Galaxy is a vital part of the development for film & television industry chain in Jing'an District, and its transformation plan is closely related to the film & television industry. For example, the original plant structure will be retained and built as a large studio for future SMG (Shanghai Media Group). Inside the park, the longest boulevard within factories in Shanghai will also become the exhibition corridor for film & television industry in the near future. In addition, since the industrial image along Wanrong Road in Daning area has already developed to a certain scale, one of the developing strategies of Jing'an Innovation Galaxy is to sign a strategic development agreement with the district government to develop audiovisual industry including film, television and audio products in accordance with the regional industrial planning. With the financial strength of leading companies such as SMG, hardware infrastructures can be shared in the park. "The film & television production industry needs heavy asset investment. The equipment for film & television shooting is expensive and fast-changing. The shared equipment helps small creative companies to reduce production costs and for leading enterprises, it can also improve the utilization rate of assets." Chen Xueqing explained.

INNOVATIVE PRACTICE OF CREATIVE PARKS

The two phases (1.0 and 2.0) Jing'an Innovation Galaxy Site Project held by Shanghai Urban Space Art Season are just the innovative practice of Jing'an Innovation Galaxy as a creative park to promote integration and interaction. "The cultural and creative industry is ultimately oriented to the C-end, that is, the crowd itself. Therefore, the park hopes to carry out industrial agglomeration and content creation, while allowing internal activities, such as providing performances, to serve the surrounding areas, meanwhile retaining the building itself is also to attract groups in cultural memory." Chen Xueqing said. With the slogan with distinctive characteristics of the past time like "We Workers Have Power", the practice case 1.0 launched in October 2017 first descriptively constructed the connection between the public and the old buildings. Take the "Art Republic" section as an example. Through collecting old factory slogans and making them into neon lights hidden in the exhibition area, people were encouraged to find and interact in various forms and old workers were attracted by memories. This creation of nostalgia allowed them to participate and interact. New art was also being exhibited as the atmosphere and connection was created. By exploring industrial materials or industrial-related themes to get close to the origins of memories, people could change the functional awareness of the site while still gaining a sense of identity in the new space. The work "Pulse" completed the metaphor of the machine age and the expression of the factory function by using a spherical device combined with the old mechanical set. "We want the public to get involved," the curatorial team representative Ma Huiting talked about their original intention, "From the factory that produces equipment to the future film & television cultural park, we avoid blunt implants, and let people understand and explore the changes in the nature of space in terms of tonality and temperament in a more gentle way."

In the practice case 2.0 launched in January 2018, the theme was further extended to "We All Have The Power", expressing a breakthrough in trying to construct in the crowd. Not just workers, new individuals were also accepted and allowed to interact with space in new ways. Changing the nostalgic theme of 1.0, 2.0 practice injected futuristic and modern design into industrial elements. For example, the second of "Zao" series uses the word "Zao" meaning "Let's on fire" instead of the word "Zao" meaning manufacturing, using the sound and photoelectric effect to stereoscopically present the meaning of the words and transformation carried out by the fonts. "In this age, everyone is a creator. The exhibition hopes that the audience is not just the viewer, but they can also participate, even become artists, and the artists can also experience the work as participants." Therefore, the special design of the circular shed in the 2.0 site project was far better than expected, which made the curatorial team pleasantly surprised. After the exhibition, the team summed up the experience, "In a semi-enclosed space such as a round shed, the artists have a particular sense of territory. So, compared to the open market, they will be more willing to take the initiative to maintain the territorial atmosphere. For those experimental performances that emphasize the sense of participation, such as 'Close Your Eyes in Darkness,' the closed-end space can also help to achieve the best results, and the audience entering the shed is less affected by external disturbances."

It is true that Shanghai does not lack the case of industrial heritage transformation and upgrading, but such practices that involve the public in transformation process are rare. In particular, the two-stage exhibition emphasizes this old-to-new transformation process so that before and after the transformation, visitors will have completely different memory points for the content. But because the space is the same, the memories can also be connected. That's the meaning of the theme "Double Exposure" – with a series of artistic transformations and resets, people can gain recognition of the industrial spirit possessed by the factory in a time-space scenario, and expand this identity to a wider range of spaces. In addition, the number of visitors was five times as much as we had expected of the two site projects. Whether it is the Jing'an Innovation Galaxy team or the curatorial team Purple Roof, the positive cooperation with each other and the consistent concept also give this site project more thoughts on the future cooperatively development methods of creative parks.

Urban regeneration with city-industry integration needs to build a sense of identity with the region's history and culture. This is the basic element for the development of the industry and the key to urban regeneration. For Jing'an Innovation Galaxy, which has established an open-minded relationship with the public before the opening, this exhibition shows that it is a creative park to show greater value or higher goals than other parks – not only to be the high-quality agglomeration of the industry, but also to serve as a spatial syntax to help improve the function of the local urban area and the public service of the community. In this way, can Jing'an Innovation Galaxy fulfill its task of revitalizing the city beyond the promotion of industrial development? The exhibition is just a starting point.

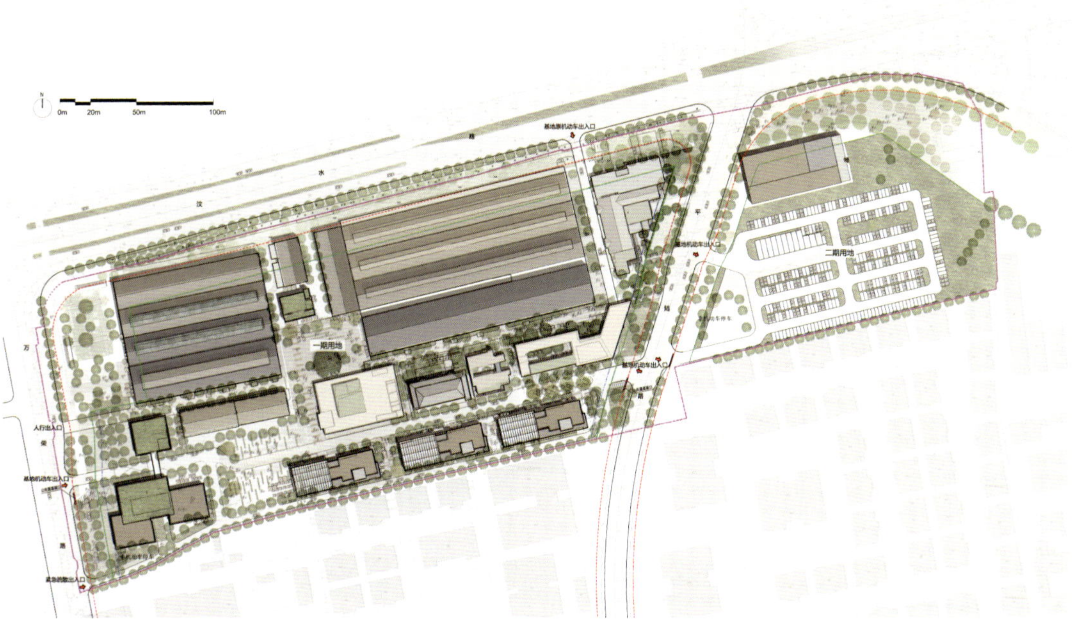

静安新业坊规划总平面图
Master Plan of Innovation Galaxy, Jing'an

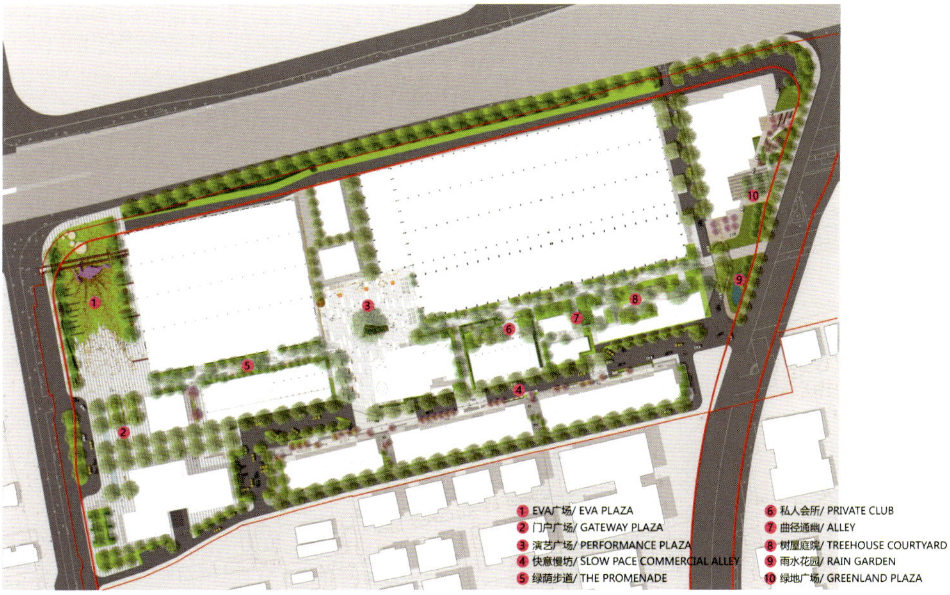

静安新业坊规划空间景观平面图
Landscape Plan of Innovation Galaxy, Jing'an

厂区中的林荫道
The boulevard in the park

彭浦工业区所剩不多的废弃工厂
A small number of abandoned factories in Pengpu Industrial Zone

“咱们工人有力量”现场俯拍全景图，兼有集市与装置
Over-the-top panorama of “ We Workers Have Power ” , with both markets and installations

《脉搏》中红色的亚克力球体，机械中喷涌而出的球体模拟血液流淌的律动
Red acrylic spheres in "Pulse", and the ball that spews out of the machine simulates the rhythm of blood flow

雕塑《新生》局部，透着历史的厚重感
Part of the sculpture "New Life", revealing the heavy sense of history

从1.0“造”到2.0“燥”：隐藏在《zao》中的齿轮和扳手是工厂生产中不可或缺的零件，也是工业革命和流水线模式的象征物
From 1.0 "Zao" to 2.0 "Zao": The gears and wrenches hidden in "ZAO" are indispensable parts of factory production and a symbol of industrial revolution and metaphor of assembly lines

小小的半圆形篷房，参与者还是闯入者？
Small semicircular tents, participants or intruders?

采访：策展团队
上海罗浮紫艺术品有限公司

两次主题的设定出于哪些考虑？

实践1.0的活动场地设置在新业坊旧厂区，目的是追忆整个工业时代，突出的重点是工人。2.0活动设置在正在改造的现场工地，本来的计划是在空间改造完成后进行，但因为工期问题，后来展览直接在工地上展开了。旧厂新生这样的改造首先是转变空间性质，从发展硬实力的冶金机械工厂到今天的影视文化园区。其实后者仍然是一个工厂，但塑造出的是精神文化，所以整个城市的发展推动者就不再局限于工人，变成"咱们人人有力量"。当时也考虑过硬性植入与影视产业有关的元素，但是后来经过团队内部讨论，决定用更巧妙的方式，采用声、光、电和新媒体元素，让整个现场的展览设计充满未来感，与1.0的怀旧复古不同，2.0让人眼前一亮，在调性上贴近园区性质，同时给予新业坊展示空间。

展览考虑了哪些受众？预期效果是否实现？

案例展的受众主要考虑的还是年轻人，但这次也针对老工人的喜好进行设计布展。出乎我们的意料，1.0和2.0的活动都吸引了很多老工人参与，很多工友是自发前来。他们找到我们，询问是否可以参加，并通过自己的微信群扩散这次活动的消息。1.0实践展开始时，他们在活动门口聚集，跟随领队，来这里聚会，并追忆过去。其实2.0设置的内容更偏向年轻人，但没有想到还是有很多老年社区居民前来，我们也有顾虑他们是否能融入，但他们其实很容易就进入展览和互动的状态。两次活动效果都远超预期，我们预计参与者最多200人，但事实上每一次都有1000多人。总体上，我们觉得第二次比第一次的效果更好，更贴合我们所希望的"反客为主，人人参与"的积极状态。

INTERVIEW WITH CURATORIAL TEAM SHANGHAI PURPLE ROOF ART GALLERY

What are the considerations for the setting of two themes?

The site for project 1.0 was set inside the old factory in Xinyefang, for the purpose of recalling the memories of the factory over the course of the entire industrial age, and the emphasis was on workers. The site for project 2.0 was the site at renovation. The original plan was in the renovated space, but due to construction schedule, the exhibition afterwards was directly launched on the construction site. Renovation as "Factory Rebirth" is to firstly transfer the spatial nature, from metallurgical machinery factory that developed hard-strength to today's film & television cultural park. The latter actually is still a factory, but what it shapes is spiritual culture. So, the promoter of city development is not limited to workers, but turns out that "we all have the power". At that time, there were also considerations about hard-embedding the elements related to the film & television industry, but after a discussion within the team, it was decided to take a subtler way, that is to make the entire exhibition design futuristic by using sound, light, electricity and new media elements. Unlike the nostalgic retro of 1.0, 2.0 gives people a bright look and is close to the nature of the park. But we also gave Xinyefang a show space.

What kind of audiences were targeted? Has the expected effect been achieved?

A:We considered young people as the main audience, but we also designed out of the preferences of old workers. Both Project 1.0 and 2.0 attracted a lot of old workers to participate in the activities, and many coworkers came spontaneously, which was a great surprise to us. They inquired us if they could participate, and spread the news of the event through WeChat. When project 1.0 began, they gathered at the entrance of the event, with the team leader, to reunite and reminisce. Project 2.0 actually set the content more towards young people, but beyond our expectation, there were still many elderly community residents coming. We had concerns about whether they could integrate, but it was actually very easy for them to enter the state of exhibition and interaction with the work. Both projects have far exceeded our expectations. We expected a maximum of 200 participants, but in fact there were more than 1,000 people each time. Overall, we feel that the second time is better than the first one, as it is closer to the positive situation we hoped that there are "changes from guest to host, and everyone is involved".

展览基本信息

主题：双重曝光 · 旧厂新生

时间：2017.10.21/2018.1.6

举办地址：静安新业坊文化创意产业基地（静安区汶水路180-210号）

主办单位：静安区人民政府

承办单位：静安区规划和土地管理局

协办单位：静安新业坊文化创意产业基地

策展团队：上海罗浮紫艺术品有限公司

策展团队介绍

上海罗浮紫艺术品有限公司

成立于2006年，是国内公共艺术领域的新生力量，也是融创意、制作、展示、交流为一体的艺术综合营运机构。作为策展团队，罗浮紫注重艺术创作与实践，崇尚“艺术即生活”的理念，为公共艺术提供战略性规划及创意执行；通过探索和创新，为艺术进入公众生活创造更多的可能性。罗浮紫自2010年起连续举办四届“静安国际雕塑展”，2007—2014年连续担任上海艺术博览会主题雕塑策展及执行单位。

文+采访

王卉[城市中国]+崔国[城市中国]

图片提供

上海罗浮紫艺术品有限公司

Basic Information

Theme: Double Exposure, Factory Rebirth

Time: 21 October, 2017 and 6 January, 2018

Venue: Jing'an Innovation Galaxy (No.180-210 Wenshui Road, Jing'an District, Shanghai)

Host: People's Government of Jing'an District, Shanghai

Organizer: Jing'an District Planning and Land Authority

Co-organizer: Jing'an Innovation Galaxy

Curatorial Team: Shanghai Purple Roof Art Gallery

Curators Team

Shanghai Purple Roof Art Gallery

Founded in 2006 as a new public art space, as well as a multi-functional institute that features creativity, production and exhibition. As a curatorial team, Purple Roof focuses on artistic creation and practice, advocates the concept of "art is life" and provides strategic planning and innovative execution of public art events. Purple Roof has successfully organized Jing'an International Sculpture Project in 2010, 2012, 2014 and 2016, and it is the key curatorial and executive team of the sculptural section of Shanghai Art Fair for eight consecutive years from 2007 to 2014.

Text

WANG Hui [Urban China] +
CUI Guo [Urban China]

Image Source

Shanghai Purple Roof Art Gallery

嘿！咱们工人有力量！

实践 1.0 中，多个带有时代烙印的标语，隐藏在展区的各个角落
In project 1.0, a few slogans with time stamps were hidden in every corner of the exhibition area

I ♥ SH
咱们人人有力量!
新业坊文化创意产业基地
新业坊
2017上海城市空间艺术季
静安新业坊实践案例展

更偏向整体未来感氛围营造的实践2.0活动
Project 2.0 was more inclined to the futuristic atmosphere

退休工人在欣赏老照片
The retired workers enjoyed seeing the old photos

退休工人的集会
The reunion of retired workers

对手绘扇面充满兴趣的年轻人
Young people who are interested in painting the covering of fans

食堂区的小朋友们捧着搪瓷饭盒，体验爷爷奶奶辈的芳华岁月
The children in the cafeteria area holding enamel lunch boxes and experienceing the old time of their grandparents

断裂与连接，滨水等城市公共空间的复兴

CONNECT THE BROKEN: RENAISSANCE OF URBAN PUBLIC SPACES, WATERFRONT AND MORE

「双向辐射」·城市更新在西岸：徐汇区西岸实践案例展

DUAL RADIATION: SITE PROJECT OF URBAN RENEWAL AT WEST BUND, XUHUI

EXHIBITION THEME: DUAL RADIATION

策展主题：「双向辐射」

项目简介

上海西岸是以上海徐汇滨江地区为载体打造的城区品牌，该区域位于徐汇区西南域，北起日晖港，南至关港，东临黄浦江，西至宛平南路—龙华港—龙吴路，紧邻徐家汇、龙华历史文化风貌区，与世博园区、后滩花园隔江相望，面积约达9.4平方千米，岸线长约11.4千米，是目前上海黄浦江两岸可成片开发面积最大的区域，区域开发总量约900万平方米。

上海西岸围绕“西岸文化走廊”品牌工程、“西岸传媒港”等核心项目，着力打造汇集国内外顶尖文化艺术、信息传媒、时尚设计、创新金融等业界领袖的国际级滨水文化金融集聚区，目标是成为与巴黎左岸、伦敦南岸比肩，独具魅力与活力的世界级滨水新城区。

策展主题阐释

西岸实践案例展借用文学写作手法“双向辐射”作为主题。双向辐射在文学写作中指环境和人互相依托、互相映衬，物中有人、人中有物，物人同化的写作手法，它的特点是自然环境与人物性格呈双向辐射关系。它强调物与人、境与人的关系，既要避免油水分离，又要防止生硬组合，牵强附会；只有结合得自然、巧妙、和谐，才能生成美感，起到深化感情和深化作品主题的作用。在西岸公共空间和公众生活的联系与互动中，观照与辐射便组成一个双向辐射的有机生长体，人、城市、空间、时间于其中都是一门艺术，它既是科学的，也是社会学的，更是一个未来的开端。

展示内容

“‘双向辐射’·城市更新在西岸：徐汇区西岸实践案例展”的主要内容为联动西岸艺术季内相关展览和公众活动，引发社会各界对滨水公共空间整治、艺术介入公共空间、历史工业遗存保护更新等问题的广泛关注，展示内容包括：星美术馆、跑道公园、油罐艺术中心、滨江公共开放空间内的十余个公共艺术作品。

在展期内，星美术馆、油罐艺术中心为在建项目，跑道公园为部分建成对外开放项目。本展将建设工地围挡为展墙，展示项目各自的设计文脉、理念、亮点以及效果，以文字、图片、多媒体等方式交织呈现。展览结合项目自身建设和城区整体有机更新，邀请公众见证并参与这一过程。同时，在西岸公共开放空间内散点呈现十余个高品质的公共艺术作品，形成一个叙事性、辐射性的公众参观路线，串联起西岸城区的时代变迁与日常生活，空间载体与文化内涵，并在西岸的整体文化语境中，使公众获得一种沉浸式、反身性的个人体验。

PROJECT INTRODUCTION

Shanghai West Bund is an urban branding and development project that aims to build upon the cultural and commercial foundation of the Xuhui Riverside. West Bund covers a total area of 9.4 square kilometers in the southwest part of the Xuhui District, including 11.4 kilometers of shoreline. The area stretches from Rihui Port in the north to Huangpu River in the east, and Wanping Rd, Longhua Rd, and Longwu Rd in the west. It neighbors the Longhua Historic Conservation Area and faces the Shanghai World Expo Site, with Houtan Park across the river. West Bund's 9.4 square kilometers of land represents the largest developable area on either side of the Huangpu River.

Shanghai West Bund is an ambitious development project that seeks to bring domestic and foreign leaders together in the fields of culture, art, media, fashion, design, and innovative finance to create a unique, world-class waterfront on par with the Left Bank in Paris and the South Bank in London.

THEME INTERPRETATION

The theme "Dual Radiation" is borrowed from literary techniques. In the context of literature, dual radiation refers to the attempt to capture the natural environment and the characters' personalities in a bi-directional radiative relationship, where both elements support, reflect and assimilate each other. It is about sublimating the story in terms of contents, aesthetics and sentiments by forging humans, objects and plots into a delicate, harmonious and integral whole, instead of leaving these features scattered and separated. Going through the history and prospect of West Bund, the idea struck us that the public space and public life of West Bund, with their connections, interactions, reflections and radiations, constitute an organically growing dual radiation, where human, city, space and time all find their artistic, scientific, sociological and futuristic expressions.

EXHIBITION

Joint efforts were made with peer exhibitions and events to raise the public consciousness of issues like management of waterfront public spaces, art in public spaces and preservation of industrial heritage. The most important exhibits included Start Museum, Runway Park, TANK Shanghai and ten pieces of public art in public spaces.

The walls surrounding Start Museum and TANK Shanghai under construction, and Runway Park just completed, were used as panels demonstrating the designs, concepts, highlights and renderings of these projects in forms of texts, graphics and multi-media. Putting individual projects in the big picture of urban renewal, this exhibition invited the public to witness and participate in this process.

As part of a joint event with other SUSAS exhibitions and cultural events, this exhibition installed ten superior pieces of public art around West Bund public spaces, guiding visitors through a narrative and radiative route, displaying the history, daily life, physical space and cultural meanings of West Bund in a holistic manner. The exhibition aimed to bring an immersive and reflexive experience to visitors in the whole cultural context of this site.

2012年9月，首届西岸音乐节在徐汇滨江绿地举办。这场为期三天的音乐节，成为徐汇滨江突破转型，发展“西岸文化走廊”的首秀。此后，西岸艺术与设计博览会、西岸建筑与当代艺术双年展等事件又为西岸贴上更多文化标签，这片位于上海黄浦江畔，与世博会址隔江相望的9平方千米空间，意欲借文化获得另一种关注。从工业区的“西外滩”到美术馆大道的“西岸”，十年布局，徐汇滨江成为上海向外的重要窗口之一，展示着这座城市雄心勃勃的未来。

魅力再发展，滨水区的内涵式改造

服务型城市最显著的特征之一就是发展理念从以生产为中心到以人为本的转变。人本关注在空间上的具体表现为从外延式开发逐渐向内涵式改造过渡。过去二十年，上海各个区域出现大规模的消费、商务空间，浦东沿岸写字楼和商品房组建的壮丽天际线诉说着上海卓越的商业成就。江景成为商业地产宣传的一环，两岸土地价值急剧抬升，黄浦江沿岸作为一种空间形态，本身的价值反被削减。相反，在西方较发达的城市，滨水区的开发建设已从单一物质空间改造向空间与功能开发并举的阶段转变，滨水区的魅力也从单纯提供绿色开敞空间，拓展为塑造城市轮廓线，乃至彰显城市多样性。

西岸的改造缘起于此。在此之前，这里曾见证民族工业的荣光，曾作为工业基地承担航运、仓储和生产等多项功能。建设于20世纪初期的日晖港货栈、龙华机场、北票码头以及上海水泥厂，是上海最早的火车站、最早的机场、曾经中国最大的水陆联运码头和最早的水泥厂。2009年，为配合世博会规划建设，后更名为南浦火车站的日晖港货栈、上海水泥厂等先后关闭或搬迁，徐汇滨江也得以在工业移除后，继续8.4公里沿江岸线的改造。

在这场浩荡的城市更新过程中，除了服务于人，如何让城市的发展记忆贮存在城市空间中并进行有机的城市更新成为改造的另一主要议题。在本次实践案例展中，三个案例都显示出对工业遗产的重视。

星美术馆原址为上海南浦火车站十八线仓库，使用面积约3077平方米。建于清光绪三十三年（1907），日晖港货栈是与沪杭甬铁路线同时兴建的两个火车站之一（另一个是上海南火车站）。作为当时全国可进行水陆转运的三个栈场之一，日晖港货栈依托黄浦江码头，从事煤炭和木柴等货物的装卸。以此历史和场域文化为元素，星美术馆的建筑师希望通过融入创办人何炬星先生追求的艺术理念，将美术馆的基础功能和艺术知识生产与建筑空间结合，建成全新的当代美术馆建筑作品。

油罐艺术中心则位于上海市徐汇滨江，改造自原中航油公司在龙华机场的储油罐、消防水池及其他配套设施。这里曾见证上海工业发展的崛起，也是上海城市发展中重要的场所记忆。1922年，龙华机场在原淞沪护军使署操场的基础上改建，因操场周围地域空旷，交通方便，且东临黄浦江，有利于陆上和水上飞机的升降，遂辟为机场，成为上海第一个陆军使用的军用机场。1929年，龙华陆军机场由航空署接管，相继开辟沪蓉、沪平（北京）、沪粤航线，并迅速发展成为东亚最大的民航机场。龙华机场在解放战争中破坏严重，于1952年整修扩建，恢复通航。直至1966年12月31日，这座建国前上海唯一的民用机场才停止了民航飞行任务，仅作为民航训练基地和航班飞机备降场。

龙华机场的储油罐内部空间开阔，挑高15米，拥有极具工业特点的穹顶以及曲面圆周空间。五个独立的油罐由“超级地面”——覆土绿化的一个新地面连接起来。“超级地面”之上是艺术公园，之下是室内展览及服务空间。保存下来的五个油罐，两个被改造为艺术展示空间，其他三个被改造为多功能展览、休闲及活动空间。油罐罐体之间，一个是适合举办大型活动、铺装石材的55米见方的广场，另一个是面积较小的草坪，更适合小型的主题活动。

跑道公园面积8.42万平方米，其前身是运营超过80年的龙华机场跑道。现存的南北向混凝土跑道由国民政府民航局建于1948年，一直服役至龙华机场飞行任务结束。景观设计公司Sasaki效仿机场跑道的动态特质，用多样化的线性空间将街道和公园组织成一个统一的跑道系统，并以水系和绿地穿插其中，将原龙华机场跑道转变为汽车、自行车、人与河流的跑道，满足人们对交通、休闲和体育运动的需要。尽管空间均是线性，但每个空间中采用了不同的材料、尺度和地形，并在道路铺装、人行通道、沿岸设计等方面进行了设计，以丰富空间体验。在为徐汇滨江地区带来舒适的现代城市公共空间的同时，跑道公园的设计希望将历史记忆融入进来，使之成为反映上海城市发展史的创意城市更新项目。

在新的城市发展背景下，这些历史空间的原本功效不可避免会被弱化，但产业遗存的独特魅力需要放大。西岸努力的方向，即是这种魅力的再发展。对此，徐汇滨江采取以下三种策略：第一，把水岸全部腾出，用作公共休憩空间。不同于过往将水岸价值最高的地段留作房地产和商业开发，西岸各个项目在启动时就以人的需求为第一要义规划水岸功能。这样的开发在短效利益的实现上或许不佳，但却有助于一座城市的长远发展。第二，西岸从徐汇衡复风貌保护区中吸取的经验是，人的生活应该是过去、现在和未来的交织。换言之，西岸原来独有的痕迹需要保留。工业遗存自有其魅力，通过新事物的积累，产生厚重的体验感。如策展人陈安达所言："我们希望在这个空间里穿梭的人，每时每刻都能感受到这个空间曾经是用来干什么的，同时这里的未来又是基于这样的空间躯干，得到新生。"第三，重构空间，将平缓且与水岸距离较远的防汛墙进行区隔，形成梯度式平台，称之为"上海Corniche"。"Corniche"一词原指戛纳的海岸线，形容道路、防汛墙到亲水平台层层递减的形态，通过梯度空间拉近人与自然、与水面的距离。作为上海的母亲河，黄浦江穿越这座城市，但因原本高筑的防汛墙，人与水的距离很远。西岸所要营建的是真正开阔、亲水的空间，借水串联城市自然和文脉，让文化和艺术这些人人可共享的内容在自然景观中展现无遗。

西岸开发，先行者的突破

事实上，这场看似充满历史必然性的发展改造仍面临多重挑战。曾经的工业基地要让位于文化活动，需要转变城市自上而下发展的观点，同时也需要资本输入逻辑的综合调整。想要实现区域开发的初衷，一套考虑到投资、规划、政策等各个层面的完备机制急需被搭建和检验。

策展人坦言，西岸改造理念从提出到实施经历了一个十分复杂的过程。比如梯度式防汛墙的设想，尽管满足了亲水需求，但这样的改建突破了原有的防汛体制。考虑到安全因素，与政府部门多次反复的沟通必不可少。诚然，除了老城区，西岸如此大范围保留工业遗存的做法极具先驱导向。因此，从构思规划到最终呈现的这个过程，项目缺少可以完整推进整项工作的实践者。陈安达总结道："规划部门重视（遗产）保留，但是重视与规范不同，缺少细节的东西。西岸的关键在于创新的担当和勇气，回归到以人为核心。"保留不同于复原，南浦车站的铁轨、油罐等的保留因涉及国有资产，需要和国家多个部门协调。这其中的困难是从上至下的执行者在开发过程中对其核心理念的坚定。西岸如今的开发成果是实践勇气的产物，对于后继的模仿者而言，建立制度体系才是解决问题的关键。在已有实践案例的基础上，制定工作执行标准，站在政府层面上具体细化方法，才能保障实践顺利完成，还原空间营建的作品意义，而非单纯政绩任务的达成。

除此之外，以文化产业为开发导向的西岸，为在中国社会环境下创建民营美术馆提供了实践样本。基于发展为上海具有文化影响力的滨江地带的目标，甚或在艺术上达到与世界艺术都会巴黎、纽约等相提并论地位的目标，西岸在规划初期就对产业进行了严格把控。把入驻行业的范围限制在信息、娱乐、设计、时尚及传媒等领域，入驻企业需要保证70%为自有发展内容，并且同意服从区域的整体规划如高度标准限制等。为了保障美术馆的纯粹性，此类文化项目不参与周边地块的连体开发。陈安达解释说，这与美术馆创建的几个必备需求相关："一是私人美术馆的创始人需要有一定数量级的藏品，否则就只是单纯的艺术中心；二是收藏者自愿将收藏展出；三是创建美术馆的本心不应带任何开发性质，如果计较与周边区域的利益关系，是开不好美术馆的。"如此，西岸现已开放的美术馆在一定程度上保障了内容和运营的质量，但是距离形成影响力仍有差距。在未来，对民营美术馆的鼓励或许需要同城市遗产保留等工作一样，需要建立可落实、可具体执行的标准。

In September 2012, the first Shanghai West Bund Music Festival was held on Xuhui Riverside greenbelt. The three-day music festival became the first show of Xuhui Riverside's transformation and developing "West Bund Cultural Corridor". Since then, events such as the West Bund Art and Design and West Bund Biennial of Architecture and Contemporary Art have labeled more on West Bund as "cultural". This 9-square-kilometer space on the banks of the Huangpu River in Shanghai, across the river from the Expo site, is intended to catch an alternative attention through culture. From the West Bund of industrial zone to the West Bund of Museum Avenue, after ten years of planning, the Xuhui Riverside has become one of the most important windows for Shanghai to show its ambitious future.

RE-DEVELOPMENT OF CHARM, THE CONNOTATIVE TRANSFORMATION OF RIVERSIDE AREA

One of the most significant features of a service-oriented city is the shift in development concept from production to people. Human-oriented attention in the space is specifically reflected in the transition from extensional development to connotative transformation. Over the past two decades, large-scale consumption and business space have emerged in various regions of Shanghai. The magnificent skyline composed of office buildings and commercial housings along the banks of Pudong showcases the achievements of Shanghai's outstanding business. The river view has become part of the propaganda for commercial real estate. The value of land on both sides of the river has been drastically elevated. However, the value of the Huangpu River riverside as a spatial form has been suppressed. On the contrary, in more developed cities in western countries, the development and construction of waterfront area have changed from a single physical spatial transformation to focusing on both space and function. The charm of the waterfront area also expands from offering green open space to shaping the contours and showing the diversity of the city.

The transformation of West Bund originated here. Prior to this, as an industrial base equipped with many functions such as shipping, warehousing and production, it witnessed the glory of the national industry. Built in the early 20th century, the Rihui Port Warehouse, Longhua Airport, Beipiao Pier, and Shanghai Cement Plant are respectively the first railway station, the first airport in Shanghai, the largest water-land intermodal pier and the first cement plant in China. In 2009, for the planning and construction of World Expo, the Rihui Port warehouse, already renamed as Nanpu Railway Station, and the Shanghai Cement Plant were either closed or relocated. Xuhui Riverside was able to continue the transformation of the 8.4-km strand after the removal of these industrial facilities. In this vast urban renewal process, in addition to serving people, how to store the city's development memories in urban space and conduct organic urban renewal has become another major issue of transformation. In this site project, all three cases have shown an emphasis on industrial heritage.

Start Museum will be rebuilt on the platform of Nanpu Station, the earliest freight train station in Shanghai, with the original architectural history and field culture as elements. TANK Shanghai, located in Xuhui Riverside at the old site of Longhua Airport, was transformed from the fuel tanks for Longhua Airport from China Aviation Oil Company and introduced various art-related functions, transforming former oil tanks into a carrier of art and life. Runway Park used to be the runway of Longhua Airport, which had been in operation for more than 80 years. And now its design emulates the dynamic characteristics of the airport runway and organizes streets and parks into a unified runway system through a variety of linear spaces.

In the context of new urban development, the original functions of these historical spaces are weakened, but in turn, the unique charm of industrial heritage is magnified. What West Bund tries to do is the re-development of this charm. In this regard, Xuhui Riverside adopts the following three strategies: First, all the riverside areas should be vacated as public leisure spaces. Different from leaving the land with the highest value in waterfront for real estate and commercial development, each project nowadays in West Bund is launched based on the needs of people, which is also the priority of planning the waterfront function. Such development may not be profitable in short terms, but will benefit the long term development of the city. Second, the experience learned from Xuhui Heng-Fu Historical Area is that people's life should be an integration of the past, the present and the future. In other words, the original unique imprints of West Bund need to be preserved, and the industrial heritage will have its own charm. Through the accumulation of new things, it will have a condensed sense of experience. As curator Chen An'da said, "We hope that people in this space can always feel what this space was once used for, and its new functions that will be given in the future based on this space structure." Third, the space will be restructured. West Bund separates the flood control wall that is gentle and far away from the water shore, forming a gradient platform called "Shanghai Corniche". The term "Corniche" originally refers to the coastline of Cannes, describing the decreasing form from roads to flood control walls and to hydrophilic platforms, and narrowing the distance from nature and water through the gradient space. As the mother river of Shanghai, the Huangpu River flows through the city. But because of the original high flood control wall, people are kept away from the water. What West Bund wants to build is a truly open and hydrophilic space. By using water to connect the city's nature and cultural context, the content that can be shared by everyone, such as culture and art, can be presented in natural landscape.

WEST BUND DEVELOPMENT, BREAKTHROUGH OF THE PIONEERS

In fact, this seemingly historically inevitable transformation still faces multiple challenges. Beyond the fact that the former industrial base paved ways to cultural events, it requires a change to the top-down view of urban development, as well as a comprehensive adjustment of capital input logic. In order to realize the original regional development, a complete set of mechanism at investment, planning, and policy levels is urgently needed to be built and tested.

The curator admits that the concept of West Bund transformation went through a very complicated process from its introduction to implementation. For example, the idea of gradient flood control walls, despite meeting the hydrophilic demand, was a challenge to the existing flood control system, so repeated communication with the government department was indispensable for safety reasons. West Bund's practice of retaining industrial remains in such a large scale is extremely pioneering apart from the old town. Therefore, from the conceptual planning to the final presentation, the project lacks practitioners who can fully lead the entire work. Chen concluded, "The planning department attaches importance to (heritage) reservations, but there is a difference between attention and norms. There's a lack of details. The key to West Bund lies in the responsibility and courage of innovation, and the core is people." Retention is different from restoration. The reservation of rails and oil tanks at Nanpu Station needs to be coordinated with various departments of the country because it involves state-owned assets. The difficulty is the top-down executors' firmness to the key concept during the whole process. The development of West Bund today is the product of practice with courage. For the subsequent imitators, the establishment of institutional system is the key to solving the problem. Based on existing practical cases, formulating implementation standards and detailing specific methods on the government level ensure the smooth completion of practice. The focus is on restoring the meaning of the works built by the space, not simply fulfilling political tasks.

In addition, West Bund, which is oriented towards the cultural industry, provides a practical sample for establishing a private art gallery in the Chinese social context. In order to be Shanghai's culturally influential riverside, or even artistically on par with world art metropolises such as Paris and New York, West Bund has strictly controlled the industry at the beginning of the planning. Industries here are restricted to the fields of information, entertainment, design, fashion and media, and companies need to ensure that 70% of production is original and agree to follow the overall planning of the area, such as height limitation. To maintain the simplicity of the museum, cultural projects of such kind do not join the development of the surrounding plots. Chen explained that this was related to several essential needs for the establishment of a museum. "First, the founders of private art museums need to have a certain number of collections, otherwise it will just be an art center; Second, collectors should voluntarily display their collections; Third, the original intentions of establishing an art museum should not include any development considerations, and it is not good to run an art museum with the attachment to the interests of the surrounding areas." In this way, the art museum opened in West Bund has guaranteed the quality of content and operation to a certain extent —but there is still a way to go from being influential. In the future, the encouragement for private art galleries may need to be implemented into executable standards, just as conducting the work of urban heritage preservation.

星美术馆设计平面图
Plan of Start Museum

白天与黑夜，西岸风光都透着水的灵动
From day till night, West Bund is glittering with water

油罐艺术中心和周围环境的奇妙融合
Harmony of TANK Shanghai and surrounding environment

油罐艺术中心俯瞰
Aerial view of TANK Shanghai

跑道公园总体平面图
Plan of Runway Park

跑道公园部分俯瞰
Aerial view of Runway Park

跑道公园纵览
Overview of Runway Park

伴着江景的跑道空间
The runway space with river view

西岸的绿地，充满活力的公共空间
The West Bund greenbelt is a vibrant public space

滑板广场成为年轻人嬉戏游玩的天堂
The skateboard square has become a paradise for young people to have fun

攀岩墙为城市居民提供了新型的公共运动设施
The climbing wall becomes a new public sport facility for urban residents

旧时龙华机场的飞机跑道，现在成为城市居民跑步休闲的好去处
The old runway of Longhua airport is now a spot for recreation

供油罐像是伫立于此的新型艺术品
The oil tanks have been tranformed to art works

公共艺术作品：《生生如环》（周力），不锈钢的环环相扣，似乎在诉说着与城市相连的故事
Public art work " Life as Circles " (Zhou Li). The interlocking of stainless steel seems to be telling the stories connected with the city

公共艺术作品：《太极》（丁乙），十字形的图案重复，细致地构造出作者心中的世界
Public art work " Tai Chi " (Ding Yi). The crisscross patterns repeat and meticulously construct the artist' s imagined world

采访：策展人陈安达

能否简单谈谈西岸的改造缘由？

西岸的改造缘起于2008年，即借着2010年上海世博会的契机开始推进沿江岸线的改造。世博会的主题是“城市让生活更美好”，那时就已经开始了城市与人的生活之间关系的探讨。徐汇滨江当时开始了第一次改造，启动了日晖港到龙耀路之间3.3千米的空间。把一些旧的工业功能空间移除之后进行改造，当时我们一直在想，到底怎样的水岸是未来人们生活需要的，是这座城市的发展所需要的。

当时确定了非常好的设计导则。第一，把水岸全部腾让出来作为公共休憩空间，在当时的水岸建设工作进程当中，这个理念是非常先进的。原来的做法通常是把水岸价值最高的地段留给房地产和商业，西岸在2008年启动时就站在了以人为本的角度去规划水岸功能。

第二，水岸除了应该具备优质的基础设施，还应该有什么？在进行了大量国际方案的征集工作后，我们发现老城是人们偏向于享受生活的地方。徐汇区有衡复风貌保护区，无论多大年龄的人到那里都会有一种特别的情怀，体会到那里的人文历史与城市之间的交融。人应该生活在过去、现在和未来当中，西岸这个地方不应该被推倒重来，而是应该保留历史独有的痕迹。所以我们当时选择将很多工业遗产保存下来进行改造，比如南浦火车站的铁轨，龙华机场的跑道、油罐，水泥厂的预均化库，总计超过40处。把一些旧的东西重新利用起来，放进新的空间里，比如把铁轨枕木找出来重新整合，做成公共休闲座椅。我们希望在这个空间里穿梭的人，每时每刻都能感受到这个空间曾经是用来干什么的，同时这里的未来又是基于这样的空间躯干，得到新生。

第三，我们把原来的空间也重新进行了划分。平缓且与水岸距离非常远的防汛墙，进行区隔后，形成梯度式的模式，即“上海Corniche”。在西岸，我们希望实现一种特别亲水的氛围，即开阔、亲水、适合公共活动，把最好的资源让给文化、艺术。坚持是难得却又必须做的一个选择，正如今年黄浦江两岸贯通工程，西岸的岸线向徐浦大桥拓展的过程中，我们依然坚持保留工业遗存、亲水性和开放性的做法。新的一年，我希望大家可以到西岸南段再体验一下“上海Corniche”。

INTERVIEW WITH CURATOR CHEN ANDA

Can you briefly talk about the reasons for the transformation of West Bund?

The project started in 2008 when the transformation of the riverbank along the Huangpu River was being planned for the 2010 Shanghai World Expo. The theme of Expo was "Better City, Better Life", so at that time, the exploration of the relationship between city and people's life had already begun. Then Xuhui Riverside began its first transformation, developing the space of 3.3 kilometers between Rihui Port and Longyao Road. After the removal of some old industrial functional space, we began to think what kind of waterfront was needed for people's life in the future, as well as for the city's development.

Some principal guidelines were decided at that time. First, the waterfront has to be completely used as public spaces. This idea was very pioneered among all the waterfront constructions at that time. Prior to that, waterfront of high value was usually left for developing real estate and commercial use. West Bund project already stood at the perspective of people-oriented planning when it kicked off in 2008.

Second, what else should West Bund be equipped with in addition to high quality infrastructures? After a few international scheme collections, we found out that old towns are usually where most people want to enjoy their lives. For instance, when people go to Heng-Fu Historical Area in Xuhui, no matter at what age they are, they all have a special feeling and experience the integration of human history and the city. Hence, people should live within the past, the present and the future. West Bund should not be wiped out and then rebuilt, instead it should retain some original and unique traces. Therefore, we chose a lot of industrial heritage to save and renovate, such as the railroad tracks of Nanpu Railway Station, runway and oil tanks of Longhua Airport, pre-homogenization warehouse of cement plant, more than 40 places in total. We reused some of the old things into new spaces, such as finding the railroad ties and re-integrating them into public resting seats. We hope that people in here can always feel what this space was once used for, and its new functions will be given in the future based on this space structure.

Third, we re-divided the original space. The flood prevention wall, which was gentle and far away from the water shore, has been separated, forming a gradient concept, that is, the "Shanghai Corniche". In West Bund, we hope to achieve a particularly hydrophilic feeling, make it open, hydrophilic and suitable for public activities, giving the best resources to culture and art. This is all what we have done in West Bund. The most difficult thing is that we have been insisting all the time. In this year's penetration project across the Huangpu River, the banks of West Bund will continue to expand to Xupu Bridge. In this process, we will still insist on retaining industrial remains, retaining hydrophobicity and openness. I hope that everyone can go to the southern section of West Bund and experience the charm of "Shanghai Corniche".

展览基本信息

主题：双向辐射

时间：2017.11—2018.1

举办地址：徐汇滨江公共开放空间（星美术馆、油罐艺术中心、西岸艺术中心等周边地区）

主办单位：上海市徐汇区人民政府

承办单位：徐汇区规划和土地管理局、上海西岸开发（集团）有限公司

协办单位：上海光启文化产业投资发展有限公司

策展人介绍

陈安达

长期致力于文化产业领域的工作，涉足文化产业项目投资、运作及管理、大型文化活动策划运营相关工作，2012年获上海市青年五四奖章。现担任上海西岸开发（集团）有限公司产业推进部部长，主要负责徐汇滨江地区文化产业项目引进和运作，大型文化活动策划运营、艺术品保税产业发展等工作。曾策划组织2013—2016上海西岸音乐节、2014—2016西岸艺术与设计博览会、西岸2013建筑与当代艺术双年展等重大品牌活动。

叶可央

毕业于同济大学城市规划专业，参与徐汇滨江开发建设近10年，主要从事区域开发的规划管理工作。现任上海西岸开发（集团）有限公司规划管理部部长，参与西岸2013建筑与当代艺术双年展策划组织、上海市街道设计导则等编制工作。

杨丹

意大利米兰理工大学建筑学硕士，同济大学建筑学硕士。曾参与组织西岸2013建筑与当代艺术双年展、2016全球水岸对话，ART WEST BUND 2016、ART WEST BUND 2017艺术季等项目。

编辑

王卉[城市中国]+崔国[城市中国]

采访

佟鑫+王卉+崔国

图片提供

上海西岸开发（集团）有限公司

Basic Information

Theme: Dual Radiation

Time:November 2017 – End January 2018

Venue: West Bund Public Open Space (Start Museum, TANK Shanghai, West Bund Art Center, and surrounding)

Host: People's Government of Xuhui District, Shanghai

Organizers: Xuhui District Planning and Land Authority Shanghai West Bund Development (Group) Co., Ltd

Co-organizer: Shanghai Guangqi Cultural Industry Investment Development Co., Ltd

Curators Team

CHEN Anda

As a veteran of cultural industry, Chen has experiences in investment, operation and management of cultural industry projects, as well as planning and operation of major cultural events. He received Shanghai Youth Award in 2012. Chen is the head of Project Promotion Department, West Bund

Group. His primary responsibilities include introducing and operating cultural industry project, planning and operating major cultural events, and promoting businesses related to duty-free artwork transaction in Riverfront, Xuhui. In his current position, he has planned and organized West Bund Music Festival (2013-2016), West Bund Art & Design Expo (2014-2016), West Bund 2013: A Biennial of Architecture and Contemporary Art, Waterfront Global Talk (2016) and Art West Bund (2016-2017). He has helped organize SUSAS 2015, Shanghai Art & Design 2016; and helped prepare the construction of landmark institutions, such as West Bund Art Center and West Bund Culture and Art Pilot Zone.

YE Keyang
After receiving his degree of urban planning from Tongji University almost a decade ago, Ye joined the enterprise of developing Riverfront, Xuhui, particularly planning and administration of area development projects. As the head of Planning and Management Department, West Bund Group, his primary responsibilities include positioning research (e.g., the functional research of Xuhui Riverfront Core Zone), documentation of area planning projects (e.g., the implementation plan of West Bund Public Open Space) and key project design (e.g., West Bund Media Port). Ye has helped plan West Bund 2013: A Biennial of Architecture and Contemporary Art and contributed to Shanghai Street Design Guidelines.

YANG Dan
Yang has master degree of architecture from both Politecnico di Milano and Tongji University. Yang has helped organize West Bund 2013: A Biennial of Architecture and Contemporary Art, Milano Animal City 2016, Waterfront Global Talk 2016, ART WEST BUND 2017, among other programs.

Edit
WANG Hui [Urban China] + CUI Guo [Urban China]

Interview
TONG Xin + WANG Hui + CUI Guo

Image Source
Shanghai West Bund Development (Group) Co., Ltd

融「汇」贯通·徐汇区T20大厦实践案例展

LINKING ZIKAWEI: SITE PROJECT OF T20 BUILDING, XUHUI

EXHIBITION THEME: LINKING ZIKAWEI

策展主题：融「汇」贯通

项目简介

T20大厦（原西亚宾馆）位于天钥桥路20号，北至肇嘉浜路，东抵天钥桥路，南接美罗大厦、汇联商厦，西邻美罗城。建筑于2015年竣工，基地用地面积0.21万平方米，建筑总面积0.24万平方米，建筑高度70米。T20项目建筑功能完备，地下为商业空间，形成地铁商场群的互通空间；地上一、二层均为公共开放空间，其中二层与六百、汇金百货、美罗城、太平洋数码等商场通过廊桥串联，结合空中绿化带和娱乐休闲设施，形成可供行人自由穿越的“空中花园”；三、四、五层为停车和设备空间；六层及以上为5A级办公空间。

策展主题阐释

徐家汇，三水交汇，融“汇”贯通，即文化之融汇、空间之贯通。如果说建设发展是一个城市外衣的更替，那么文化发展就是城市内核的演变，两者相辅相成，互为依托。随着新旧建筑相交相融，徐家汇的更新所带来的空间多样性，以及未来的可变性将令人期待。

展示内容

本次实践案例展选择与百姓生活密切相关的T20二层半开放公共空间，从城市建设发展、艺术文化发展的实际出发，将作品展览与系列文化活动的开展等多种形式相结合，展出徐家汇城市空间若干实践案例和不同领域艺术家的作品，吸引市民关注城市空间、艺术人文的发展和变化，共同提升城市活力和精神品质。策展团队在展示空间内做了以内方外圆为雏形的极简分割，形成由内而外、相融相益的格局。徐家汇实践案例以图片、文字、模型、视频等形式有机地立于场地外圈，面向都市，是实例与实际间的对话；艺术家作品作为“文化”的缩影、形而上的文明彰显，置于场地内圈，作品选取上的多元视角又折射出异曲同工的文化内涵。

PROJECT INTRODUCTION

Completed in 2015, T20 Building (formerly West Asia Hotel) is located at No. 20 Tianyaoqiao Road, bordering Zhaojiabang Road to the north, Tianyaoqiao Road to the east, Metro Tower and Huilian Mall to the south, and Metrocity to the west. It covers 0.21 hectare, with 2,400 m^2 of building area and 70 m of height. The building contains a variety of functions. The underground retail space constitutes part of the Metro Station Retail Square. The first and second floors are public spaces, and the second floor is connected to Shanghai No.6 Department Store, Huijin Department Store, Metro City and Pacific Digital Plaza via "covered bridges", which further form passable "Sky Garden" with green belts and entertainment facilities above ground. The third to fifth floors are for parking and service equipments. The sixth and upper floors accommodate 5A office spaces.

THEME INTERPRETATION

Xujiahui is at the intersection of three rivers, and what's more, of cultures and spaces. If construction works transform the appearance of a city, then cultural development is the core of its evolution. Either is the corroboration and background of the other. As new projects and old buildings mingle, Xujiahui is expecting a myriad of spaces and a future of variability.

EXHIBITION

This site project was held on the second floor of T20 Building, a semi-public space that closely relates to people's everyday life. Considering the existing conditions of urban construction and cultural development, this project adopted exhibition and serial events to showcase a number of site specific projects in Xujiahui urban space, and art works composed by a variety of creators, to attract public attention to urban space, art and humanistic development, also to make the city a more dynamic and elegant place. The curatorial team arranged the exhibition space based on the minimalist prototype of "excircle of square", creating a harmonious and penetrative layout. The project established a dialogue with the real site via pictures, documents, models and videos of the site placed around the outer circle and facing the city. Artistic works were placed in the inner circle, as the epitome of transcendental "Culture" and "Civilization". The diversified perspectives of these works reflected the irreducible and yet fundamentally harmonious heterogeneity of cultures. Art enlightened the city, and was also accommodated by the city.

T20大厦的前身，是20世纪末徐家汇的地标建筑——西亚宾馆。宾馆由分别建造于1979年和1995年的主、附楼组成。二十多年过去了，徐家汇作为上海城市中心的地位愈发显著，徐家汇商圈也随之不断更迭，外部环境的巨变和其本身结构的内在问题最终让这座历史性建筑落入被淘汰与改造之境。于是，超甲级写字楼T20大厦作为徐家汇城市更新的重要一环拔地而起。

这样的建筑改建，只是上海这座大都市生长与更新的一角。在徐家汇，这样的变化在未来注定会不断发生，不仅限于一座建筑，更是面对社会环境巨变的整个区域转型——以商业为阶崛起的徐家汇在未来的改变。

徐家汇商业：
从一夜繁荣到内忧外患

徐家汇的商业形成开始于“德昌顺南货店”——徐家汇第一家商业意义上的店铺。20世纪初，由于外国资本的介入和基础设施的改善（例如租界内自来水的供应、有轨电车线路的延伸等），徐家汇商市渐起。根据《徐汇区志》，至30年代中期，散布在徐汇区的私营和个体商店已有近700家，这些店铺主要集中在徐家汇和今淮海中路、乌鲁木齐中路一带（如永茂德陶瓷店、同益祥仁记绸布庄、裕兴皮鞋店等）。建国后，中国百货公司上海分公司第六门市部在徐家汇开业，成为徐家汇第一家百货公司。改革开放后，徐家汇商业迎来重大机遇，1986年《上海市城市总体规划方案》中，徐家汇被列入城市副中心进行规划建设，四年后，徐家汇商城的建设标志着徐家汇商圈正式起步。商业中心设计之初提出的设计理念是，让商业中心从中心广场向四周辐射，形成开放的空间。

某种程度上，徐家汇可以被称为上海成长最快的商圈之一。从规划开始到一跃成为与南京路和淮海路比肩的上海第三大商业中心，徐家汇的发展用了不到十年。这和徐家汇惯行错位竞争的发展理念以及数码市场的专业支撑密不可分。借助地铁一号线和闵行区住宅兴建的契机，太平洋百货、东方商厦和第六百货先后建成。紧随其后，美罗城和港汇恒隆广场作为徐家汇二期工程的落成也丰富了徐家汇商圈布局，形成了更具兼容性的完整商圈体系：东方商厦以吸引中高收入顾客为主；太平洋百货将重点放在关注流行时尚的女青年上；第六百货的经营走薄利多销的实惠线路；港汇恒隆广场则更注重奢侈品牌的入驻，如上海首家Burberry店铺就开设在这里。与此同时，太平洋数码一、二期和百脑汇三巨头在徐家汇形成了大家口中的“数码王国”，电脑和相关配件的交易量一度占据整个上海市场的近60%。截止到2010年，徐家汇商业建筑面积达88万平方米，日均客流量70万人次，年度商品销售额超150亿元。此后，徐家汇的发展逐步突破原来的六大商场，开始向周边辐射，形成包括交通大学在内的九大核心的多点式商圈形态——徐家汇“汇”聚了更广阔的天地。

纵然如此，徐家汇的商业发展在面对互联网驱动下的经济变革时仍显脆弱。徐家汇社会零售销量连续下降已经是不争的事实。在越来越多外来客的眼中，徐家汇已经不再是上海的核心地标。一方面，电商的兴起使“买电脑去徐家汇”这一口号成为历史，数码市场在规划中被减少，导致徐家汇最大的零售特色逐渐模糊；另一方面，面对市场环境迅速变化的现实，徐家汇却还保持着老派的零售业态，曾经错位经营的优势逐渐消失。原本的交通优势也不再明朗，如今3条轨道线路交汇于此，20多个出口分散于四周，加上更为复杂的地面交通系统，徐家汇中心区域在提供便捷的购物通道方面遇到了巨大的难题。

从消费者角度究其原因，除了电商带来的价格冲击，更多的原因是消费观念的变化。消费已不仅仅是商品的购买行为，还涉及其背后的文化内涵，以及品牌是否能传递个人价值。相应地，购物空间也不仅仅是物质商品的消费场所，还是一个可以获得深度体验的休闲地带。显然，徐家汇现有商业体无法满足这种高期待和高要求。而那些通过借助这种变化为消费者提供更丰富的商业体验，提出新鲜商业概念的商场和商圈则成功突围。例如，2013年建成的位于淮海路商圈的上海IAPM，其夜间购物概念一度“重塑”了淮海商圈的生态系统；同年开业的静安嘉里中心，也因细分自身定位为轻奢品牌而在南京西路商圈中找到新的突破口。徐家汇商圈在市区级商圈竞争中逐渐失势，在越来越多刚刚兴建的社区型商业综合体面前，“别无二致”的徐家汇没能留住它原本的客流量。

T20大厦实践：
一种发展可能

如火如荼的商圈竞争和消费者对于文化内涵的追求让在地内容的开发和引入变得至关重要。换句话说，徐家汇此时的商业亟须深挖文化内涵，从而获得持续生长的养料。早于商业发迹的两百年前，是文化特色让徐家汇在整个上海区域先行一步。徐家汇之徐家，意指徐光启之后人在这里为徐光启结庐守墓并将这里兴建成镇。而徐光启这位率先开展农业实验的明代科学家、思想家，让徐家汇之“汇”超越李从泾、肇嘉浜之流相汇的地理意义，变成富有文化意义的中西文化交流之所。

2017年颁布的徐家汇地区“十三五”规划中，文、商、旅的联动发展以及文化、产业、交通和环境四大功能的统筹发展被作为重点战略提出，如何重新借助文化迎来徐家汇新一轮的发展成为现今重要的课题之一。某种层面上，T20城市空间实践案例展为徐家汇区域的更新提供了一个参考方向。首先，如开篇提到的，T20大厦改建本身，对于原有建筑的拆除重建，是对不适应周边环境风貌和发展要求的城市空间进行的调整，这是区域产业功能转型的基础硬件载体，也是更新所需的基础条件之一。其次，T20大厦改建作为“徐家汇空中连廊”项目的一部分，代表了徐家汇试图改善交通功能的诉求，尤其是针对步行体验差进而影响到购物体验这一问题进行的深入探讨。“空中连廊”项目希望通过搭建二层平台的方式将徐家汇中心区域，包括商场和写字楼在内的场所用物理方式连接起来。这样的连接方式可以提升区域内行人的通达效率，将行人与车流分离，也会缓解徐家汇中心区域地面交通复杂这一情况。换句话说，连廊的建造会帮助改善整个区域快行、慢行交通系统；而且因为提供了进入建筑的另一种方式，连廊间接有助于楼宇商业价值的提升。最重要的是，作为商业大厦的T20开放地面一、二两层，将其作为公共空间使用，打破了原本的空间边界，将一部分商业功能置换成文化功能。这是商业与公共文化的融合，也是商业对城市公共生活和城市更新的“反哺”。这一方案也和空中连廊呼应：一是连廊平台的打造天然形成开放空间；二是连廊和T20开放空间会产生一种相互作用——平台提供的便捷通路会吸引更多人来到T20的公共空间，同时，借助连廊在空间上产生的延伸，空间和行人互动的辐射范围和影响也随之扩大。

这些基础设施的完善最终为挖掘徐家汇文化以激发区域活力而服务。2017城市空间艺术季的T20艺术展览所要传达的，是通过艺术引导人群发现并关注徐家汇商业背后那些古今中外文化交流的历史，和其对这个区域乃至整个城市产生的影响。在主策展人傅春禹的规划中，“融汇贯通”的主题体现的是徐家汇独特的文化和空间要素。“融汇”是指文化的融汇，即启蒙之光、海派之源和时尚之汇，而“贯通”则是对未来连廊建成的期待。在展览主题的设置上，傅春禹通过展出一些融合中西元素和技艺的作品，来传达徐家汇一直所传承的“文化融汇”风格。例如，《仙人来系列》（杨冬白）通过现代工业的切割铝板来诠释山水造型，把中西方审美概念结合起来，用工业的加工方式连接中国传统艺术；《中国戏曲》（张海平）运用现代雕塑语言和着色，来诠释中国传统戏曲人物的亮相造型。相比于徐家汇是上海众商圈之一的表面印象，其蕴藏的丰富文化，才是让徐家汇更有魅力的关键所在。

包括这次展览在内，徐家汇近些年在如何塑造城市文化并将其作为区域核心竞争力方面做了很多尝试。如“徐家汇源”景区，它以“海派文化策源地”为口号，将分散在徐家汇各处的重要历史建筑有机串联起来。这种独特的开放型都市景区，在上海也是首家。徐家汇如今的确面临发展瓶颈，而在这个突破过程中，在地理区位、交通条件和人文资源上都有明显优势的徐家汇仍充满机会。

The former life of T20 Building was the West Asia Hotel, a landmark in Xujiahui in the late 20th century. The main building and the annex building of the hotel were built in 1979 and 1995 respectively. Over the past two decades, Xujiahui has become a prominent city center in Shanghai, and the Xujiahui commercial zone has also undergone drastic changes, externally and inherently, which has finally made this historic building fall into the fate of being transformed. As a result, T20, a super-grade office building, becomes an important part of the urban renovation of Xujiahui.

T20 building transformation is merely a glimpse of the growth and renovation of Shanghai. In Xujiahui, facing the change of the social environment, more changes of such are expected to happen in the future, not only a building but also the entire district. The rise of Xujiahui based on business will definitely bring changes.

COMMERCE IN XUJIAHUI: FROM PROSPERITY TO PREDICAMENT

Dechangshun Goods Store is regarded as the first commercial store in Xujiahui. At the beginning of the 20th century, with the involvement of foreign capital and the improvement of infrastructure (such as the supply of tap water in the concessions, the extension of tram lines), Xujiahui commercial market began to rise. According to *History of Xuhui District*, there were nearly 700 private-owned shops in Xuhui District by the mid-1930s. They were mainly located in Xujiahui and the current Middle Huaihai and Middle Wulumuqi roads (such as Yongmaode Ceramics Store, Tongyixiang Ren's Silk & Cloth and Yuxing Leather Shoes Store). After the founding of the PRC, Shanghai branch of the Chinese Department Store opened its sixth store in Xujiahui, becoming the first department store in the district. With the Reform and Opening, Xujiahui commercial zone seized the opportunity. In the Shanghai Comprehensive Master Plan (1986), Xujiahui was planned to be developed as the city's sub-center. Four years later, the construction of Xujiahui Mall marked the start of the Xujiahui commercial zone. The design concept at that time was to form open spaces from the central square outwards.

To a certain extent, the growth of Xujiahui commercial zone was the most rapid in Shanghai. From the planning stage to becoming the third-largest commercial center in Shanghai after Nanjing Road and Huaihai Road, Xujiahui reached such a peak in less than ten years. It is because of the dislocation competition strategy and the support of the professional digital market. Additionally, with the open of Metro Line 1 and the development of Minhang residential areas, Pacific Department Store, Oriental Shopping Center and Shanghai No.6 Department Store were successively completed, followed by the Metro City and Grand Gateway Shanghai. By then Xujiahui commercial zone had formed a more compatible and complete business structure: the Oriental Shopping Center attracts middle and high income customers, Pacific Department Store focuses on trendy young communities, Shanghai No.6 Department Store operates an affordable marketing strategy with small profits but quick turnover, and Grand Gateway Shanghai pays more attention to luxury brands, for example, the first Burberry store in Shanghai was opened there. Meanwhile, because of the existence of three digital markets, namely Pacific Digital I, Pacific Digital II and Buy Now, Xujiahui became known as a "digital kingdom", where the trading volume of computers and accessories used to make up to nearly 60% of the entire Shanghai digital market (Source: Shanghai Morning Post). In total, the commercial construction area of Xujiahui reached 880,000 square meters in 2010, with an average daily flow of 700,000 people and annual retail sales exceeding 15 billion Yuan. Since then, Xujiahui has started to extend beyond the original six shopping malls, and cover a wider area - including Shanghai Jiao Tong University - to form a multi-point business model.

However, Xujiahui's commercial development is still fragile in the face of economic changes driven by the Internet. It is a fact that Xujiahui's retail sales have been declining continuously in the past few years. In the eyes of visitors, Xujiahui is no longer a landmark. The rise of e-commerce has made the slogan "Buy a computer from Xujiahui" seem senseless. With the reduction of digital businesses in the development plan, Xujiahui has lost its distinctive retail feature. On the other hand, despite the rapid changes in the market, Xujiahui still maintains its traditional retail formats and lost its advantage of dislocation competition. The advantage in traffic is also at risk - with three metro lines meet here and more than 20 exits around, in addition to a more complicated ground transportation system, the central area of Xujiahui is encountering a huge traffic difficulty.

From the perspective of consumers, besides the price disadvantage caused by the e-commerce retails, it is the change of consumption concept. Consumption is not just about purchasing goods but rather more about the cultural implications, about whether or not a brand fits and expresses individuals' attitudes and values. Correspondingly, consumption space is not only a place of physical commodities, but also a leisure zone where customers could have a profound experience. Xujiahui's existing commercial spaces apparently fail to meet these new and higher expectations and requirements. Those shopping malls and commercial zones with new consumption concepts gain the opportunity to break through. For example, IAPM Mall (built in 2013) reshaped the retail ecosystem of Huaihai Road commercial zone with its night shopping idea. Jing An Kerry Center, opened in the same year in West Nanjing Road shopping district, found itself a niche market. In contrast, Xujiahui commercial zone has gradually lost its advantages in the municipal-level competition of commercial zones. With the emergence of more community-based commercial complexes, Xujiahui is losing its original customer flow.

T20 BUILDING PRACTICE: A POSSIBILITY FOR DEVELOPMENT

The fierce competition among different commercial zones and the consumers' pursuit of cultural implications make it crucial to introduce in-site content to the area development. In other words, the growth of Xujiahui commercial zone needs to source from the local culture to obtain chances at this moment. Two centuries before the commercial development, it was a culture that made Xujiahui develop ahead of the rest of Shanghai. It is noticeable that in the Chinese context, Xujiahui originally was the place where the descendants of "Xujia" (Xu family) stayed to protect the tomb of their ancestor XU Guangqi and then gradually built into a town. XU Guangqi was a scientist and a thinker who took the lead in carrying out agricultural experiments. He makes the word "Hui" transcend its geographical significance - the convergence of Licongjing Creek and Zhaojiabang Creek, and develop a cultural significance, representing the exchange between Chinese and western cultures.

In the 13th Five-Year Plan of Xujiahui district, the joint development of culture, business and tourism, and the coordinated development of four functions (culture, industry, transportation and environment) were proposed as key strategies. Culture has become one of the important topics for Xujiahui's further growth. To some extent, T20 site project has provided a reference for this direction. First of all, as mentioned in the opening section, the demolition of the original hotel building and the reconstruction is an adjustment move for the urban spaces that no longer satisfy the surrounding development requirements. This is a basic hardware carrier for the transformation of regional functions, and also one of the basic requirements for urban regeneration. Secondly, as part of the Xujiahui Skywalk project, T20 Building represents Xujiahui's demand for improving traffic conditions; especially the poor walking experience has adversely affected people's shopping experiences. Xujiahui Skywalk project aims to connect the shopping malls and office buildings in the central area of Xujiahui by building elevated platforms. Such a connection system can improve the accessibility to the area, keep pedestrians away from the traffic flow, and alleviate the complex situation of ground transportation in the central area of Xujiahui. The construction of the Skywalk will fairly improve both the motorized and non-motorized traffic system in the whole district. Furthermore, easier access also means it will indirectly contribute to the increase of commercial values of buildings. Most importantly, T20 Building opens its first two ground levels as public spaces, replacing part of the commercial function with cultural and public functions. This is an integration of business and public cultures and a way to feed the public life in return from the business. This site project echoes the Skywalk project as well. First, the creation of the corridor platform will naturally form an open space. Second, the corridor and the open spaces in T20 Building will create an interaction - easy access provided by the corridor will attract more people to T20, while with the extension of the space created by the corridor, the variety and attractiveness of the interaction between the space and pedestrians will expand accordingly.

The improvement of infrastructures mentioned above will ultimately serve the cultural exploration of Xujiahui to vitalise the area. The purpose of the T20 site project is to exhibit the past and present cultural exchanges between China and the West, and its influence on the area and the whole city. According to the curator FU Chunyu, the exhibition theme "Linking Zikawei" means "Integration and Comprehension", which reflect the distinctive cultural and spatial elements of Xujiahui. "Integration" refers to the convergence of culture - the light of enlightenment, the source of the Shanghai-style and the center of fashion, and "Comprehension" is the expectation for the forthcoming skywalk. The works included in the exhibition combine both Chinese and Western features, showing integration has always been a tradition in Xujiahui. For example, the Xianrenlai Series (YANG Dongbai) interprets the traditional Chinese landscape painting through the aluminum extrusion technique of modern industry, combining Chinese and Western aesthetic concepts and connecting traditional art with industrial techniques. Chinese Opera (ZHANG Haiping) uses artistic sculptural language and coloring to demonstrate the appearance of Chinese traditional opera characters. Compared with the common stereotype of Xujiahui as just a commercial zone in Shanghai, understanding the culture of the site makes Xujiahui more appealing.

Xujiahui has made many other such efforts in recent years to shape the culture and employ it as the core competitiveness of the district. For example, embedded with the slogan of "Culture Source of Shanghai-Style", "Xujiahui Origin" tourism zone organically connects the important historical buildings scattered throughout the district. It is the first open metropolitan tourism area in Shanghai. Admittedly, Xujiahui has reached its development bottleneck. However, in the effort to break through, it's still full of opportunities with its apparent advantages in geographical location, traffic conditions and cultural resources.

T20大厦建于西亚宾馆旧址
The site of the T20 Building used to be West Asia Hotel

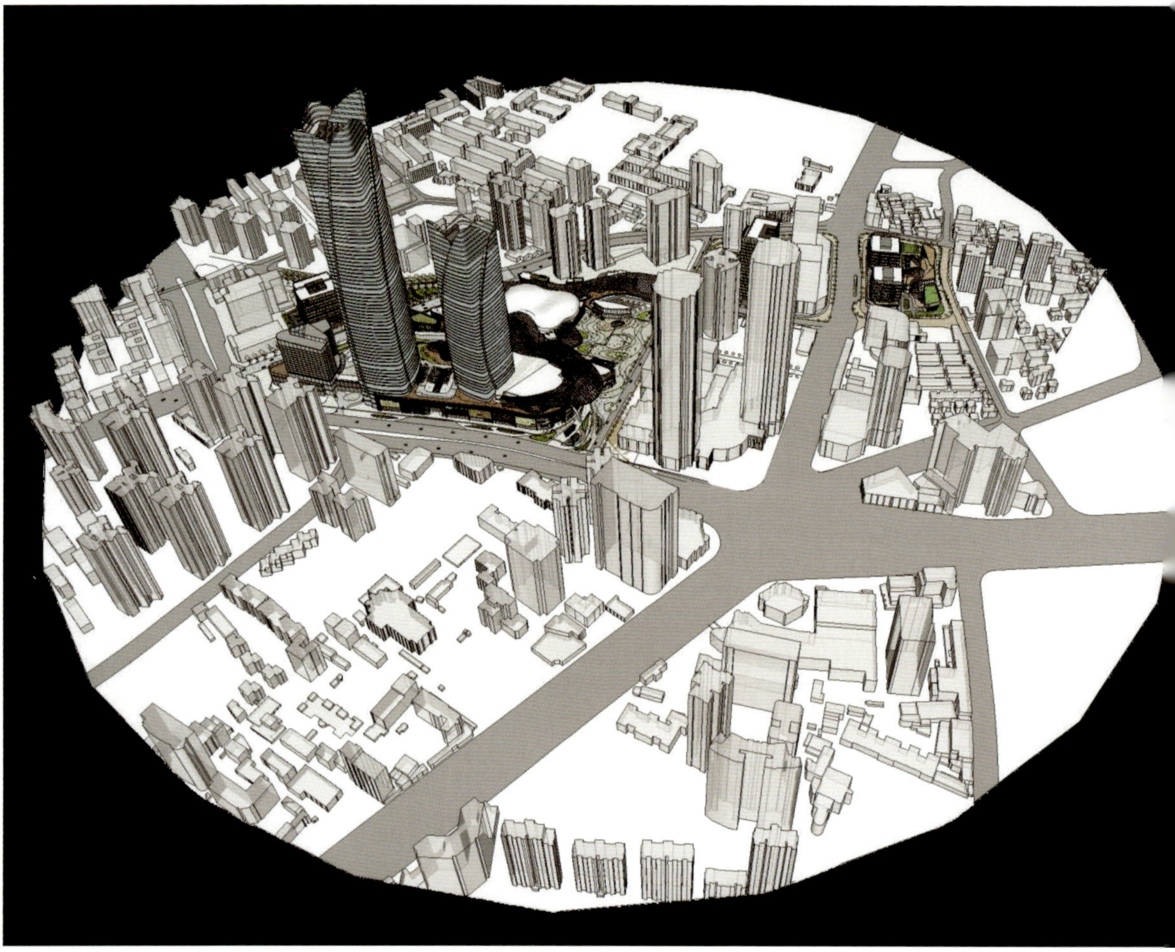

《徐家汇空间模型》中为市民展示的未来城市空间格局
Xujiahui Space Model shows the future urban spatial pattern of Xujiahui and its surrounding area

《仙人来》（杨冬白）以现代工业切割的铝板，辅之艺术家的灵动触媒，营造了一个如梦如幻的仙境
Xianrenlai (YANG Dongbai) uses aluminum plates cut with modern industrial techniques and the artist's innate sense to create a dreamlike fairyland

《中国戏曲》（张海平）用现代雕塑语言和着色手法，重释中国传统戏剧人物的演绎状态
Chinese Opera (ZHANG Haiping) reinterprets the appearances of traditional Chinese opera figures with modern sculptural language and coloring

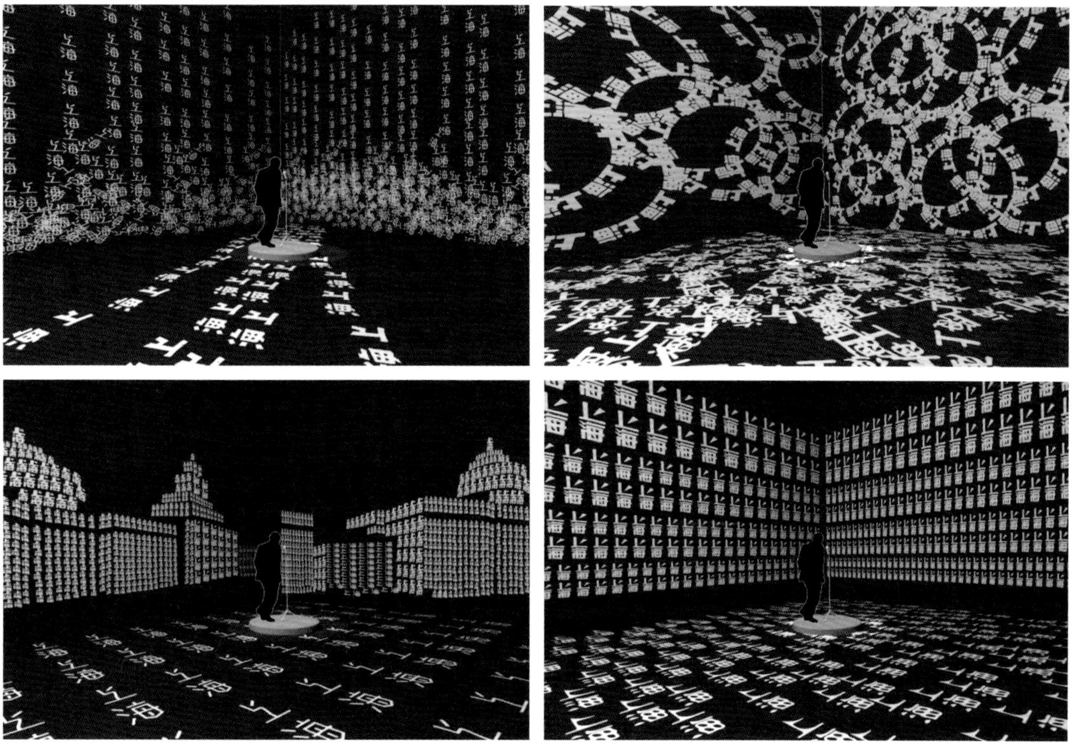

《城市，声音，文字》（蒋飞、王静艳）试图以声音唤醒对一座城市历史的记忆，其中所有“上海”字体均来自老上海的印刷品
City, Sound and Character (*JIANG Fei and WANG Jingyan*) *awakes urban memories with sounds. All the characters shown came from old prints*

采访：策展人傅春禹

整个展览从设计、布置到最后的呈现都有哪些挑战？

布展上主要是两方面的限制：时间和空间。时间上因为比较匆忙，本来考虑邀请艺术家以“融汇贯通”为主题进行大型创作，因没有足够的准备、构思和策划的时间，只能转而选择邀请现有的、与主题相关的作品进行展示。空间上的局限主要是布展地T20大厦业主方面对于空间有很多限制条件，可以改造的余地很小，限制了很多布展设想。包括之后在宣传方面也是，缺乏部分大厦业主方的配合，使得展览整体呈现效果没有达到预期目标。但好在和天华的配合过程中基本克服了大部分的困难，也得益于作品本身的艺术品质的支撑。

在作品的选取上主要有哪些考量呢？

我们的主旨是“融汇贯通”，其中这个“汇”是一个双关，既指徐家汇，也指“融汇”这个概念本身。策展时联想到了一次自己的观展经历——关于贵霜帝国将西方文化、宗教传播到东方并且包容并蓄的一些历史，让我将其和徐家汇城市中心联系到一起。20世纪初的徐家汇也是这样一个角色，比如土山湾，1913年引进彩绘玻璃并烧制成功，1915年带着木雕牌楼和木塔组合参加旧金山世博会获奖，这是中西文化兼容并蓄的见证，也是跨地域性文化积淀的魅力。这里面涉及时间和空间两个概念，所以出于对时空融“汇”贯通串联的考虑，就有了我们看到的很多结合中西方技艺、元素的内容和作品。

选取作品时我们没有过多考虑徐家汇地理位置这个点，而是更愿意拓展到徐家汇之外的思考范畴，希望借此引发对于城市空间的拓展思考和探讨。因为城市的构建除了硬件之外，真正的核心还是我们称之为文化的软实力。所以在大的设定上，我们的布局也是将体现城市发展的展板向着整个城市，通过内容来映射整个城市空间；而相关艺术作品放在内部，因为它们代表的是文化这一城市内核。

有考虑和徐家汇这边空中连廊系统做一个关联设计吗？

考虑过，展览中本身就有连廊案例，做成艺术展览的方式也是希望改变展板式的案例展现方式，让人们能切身体会这个空间。但是因为施工时间没有办法契合，开展前连廊还没有开建，到最后，连廊的建设因为挡住了其中一个入口通路反而影响到了展览。

原本展览方案中还有哪些方式的呈现呢？

一个是大型装置，本来是希望在电梯入口平台做一个悬挂式的装置作品。打算使用城市中被拆卸的老旧门板做一个螺旋塔状的东西：外面是模板，里面是镜面。内部的光线被反射后形成的效果会给参观者带来一种迷失感，模拟身处城市的感觉。另外希望有一个互动娱乐的教学内容，本来打算邀请玻璃艺术家在这里做一个关于手绘玻璃叠加的艺术工作坊，也是这个空间之后可以考虑的一种利用方式。但是后来因为展览日程在冬天，空间又是半开放，整个环境温度太低而没有实现。

对于城市空间，您个人的理解是怎样的？

城市空间的范围可以无限扩大，从单个建筑到整个城市，我们需要探索的是每一件城市空间作品建造背后的内容。这些空间作品带给我们什么？可能是正面的，也可能是负面的，因为很多东西需要通过时间或者历史来检验其正确与否。而对于这些与公共空间有关的工程的问题，艺术可以成为一种很好的探讨手段或者介质，因为它拥有一个相对中立的考量角度。我会对城市拆迁和老建筑的保护有更多关注，在创作和这些内容相关的作品时，更愿意从艺术公共性的角度去解构和重塑空间。

INTERVIEW WITH CURATOR FU CHUNYU

What were the challenges of the exhibition?

They came from two aspects, time and space. At first, I thought about inviting artists to create large-scale works based on the theme "Linking Zikawei", which means "Integration and Comprehension", but had to give up the idea due to tight schedule and chose existing works instead. The limitation regarding space was that we couldn't do many transformations to the venue in T20 because the owners had strict restrictions. Some of them were not very supportive for the promotion either. The overall effect of the exhibition was not satisfactory as expected. Fortunately, most of the difficulties were overcome in the process of cooperation with Tianhua, which also benefited from the artistic quality of the work itself.

How did you choose the exhibits?

The character Hui in our main theme has a double meaning, referring to both Xujiahui and the occurrence of integration. I once went to an exhibition about the history of Kushan Empire that brought Western culture and religion to the East. This personal experience inspired me because Xujiahui played a similar role as such in the early 20th century. For example, T'ou-Sè-Wè (Tushanwan), where generations of Chinese artisans were trained, had successfully learned glass staining techniques from the west and produced by themselves in 1913. In 1915, T'ou-Sè-Wè made a carved wooden arch and a wooden tower and won at the San Francisco World Expo. T'ou-Sè-Wè witnessed the integration of Chinese and Western cultures, and also, it showed the charm of trans-regional cultural accumulation. In concept, it is a design of time and space, so our main goal is to link these two factors together. The result is what we saw at the exhibition.

Honestly, we did not consider too much about the geographical meaning of Xujiahui when selecting works, as we expected to provoke a discussion about urban space beyond Xujiahui. According to the construction of a city, in addition to bettering physical quality, the real core is culture. The layout of the exhibition reflected this idea, with the broad peripheral signifies as the city, and the works were put inside, representing the core of the city, i.e. culture.

Have you considered creating something in relation to the Xujiahui Skywalk project?

The skywalk project was actually included in the exhibition. It was a change from the usual way of exhibition boards in order to let visitors really immerse themselves in the space. But the timing of construction didn't quite fit. By the time of the opening, the project had not kicked off. And when it did happen, it blocked one of the entrances to the exhibition.

Is there any other exhibition ideas in the original exhibition plan?

One is a large installation. We conceived a hanging installation on the elevator entrance platform, using abandoned old doors to make a spiral tower-like object. The outside would be wooden plate, and the inside be mirrors. The internal light reflected would give visitors a sense of loss, just as you would feel in the city. Another idea was to add interactive educational content, for instance, a hand-painted glass workshop, which would also be a possible use for this space after the exhibition. This didn't happen as the exhibition was in winter, and the space was half open; the temperature was too low to do this.

What is your understanding of urban space?

The scope of urban space can infinitely expand. From the construction of a single building to the entire city, what we need to explore is the implication of each space. It can be positive or negative because only time can tell. As for the engineering issues related to public space, art would be a great medium for discussion, because it holds a relatively neutral viewpoint. I've been focusing more on urban demolition and protection of old buildings, and I tend to deconstruct and reshape the concept from the art publicity perspective in my work regarding these topics.

展览基本信息

主题：融“汇”贯通

时间：2017.12.10—2018.1

举办地址：天钥桥路20号（T20大厦）

主办单位：徐汇区人民政府

承办单位：徐汇区规划和土地管理局

协办单位：T20大厦（原西亚宾馆）

策展团队介绍

傅春禹

上海大学美术学院美术学博士，中国美术家协会会员、中国雕塑学会会员、上海美术家协会会员，现任上海视觉艺术学院美术学院雕塑系讲师。作品曾被中国美术馆、北京国际双年展组委会、瑞士银行等单位机构收藏，并多次入选国家、省部级各项美术大展。多篇学术论文发表于南大CSSCI、北大中文核心期刊。独立主持上海市文化艺术发展基金艺术创作项目，参与教育部人文社会科学研究青年基金项目。

上海天华城市规划设计有限公司

天华集团专注于城市建设领域的专业服务，是拥有“创意、科技、管理”三大业务板块的相关多元化产业集团。天华在不断完善业务板块的同时，致力于促进文化与城乡建设的融合。2013年参与策展港深城市\建筑双城双年展“边缘居住”板块，2015年参与SUSAS上海设计中心主题展，2016年参与上海城市设计挑战赛，2017年举办珠海民宿设计大赛。

文+采访

王卉[城市中国]+崔国[城市中国]

Basic Information

Theme: Linking Zikawei

Time: December 10, 2017 – January 2018

Venue: No. 20 Tianyaoqiao Road

Host: People's Government of Xuhui District, Shanghai

Organizer: Xuhui District Planning and Land Authority

Co-organizer: T20 Building (former West Asia Hotel)

Curators

FU Chunyu

Ph.D. in Fine Arts from the School of Fine Arts of Shanghai University, a member of the Chinese Artists Association, China Sculpture Institute and Shanghai Artists Association. He is currently a lecturer in the Sculpture Department of Shanghai Institute of Visual Arts. His works have been collected by the China Art Museum, the Beijing International Biennial Organizing Committee, the Bank of Switzerland and other institutions, and he has been selected for various national, provincial and ministerial art exhibitions. He has published several academic papers in CSSCI and Peking University Chinese core journals. He has independently hosted the art creation project of Shanghai Culture and Art Development Fund, and participated in the Humanities and Social Science Research Youth Fund Project of the Ministry of Education.

Shanghai Tianhua Urban Planning & Design Co., Ltd

Tianhua is a group operating in a variety of practices, including innovation, technology, management, in particular professional services in urban construction. While strengthening its existing businesses, Tianhua is dedicated to promoting the convergence of cities and countryside. Tianhua helped curate the Marginal Living section of Bi-city Biennale of Urbanism\Architecture 2013, participated in the SUSAS 2015 theme exhibition in Shanghai Design Centre, Shanghai Urban Design Challenge (2016), and hosted Zhuhai B&B Design Competition.

Text

WANG Hui [Urban China] + CUI Guo [Urban China]

Image Source

FU Chunyu

《蚀年》（杨腾）用一种孩童般调侃的方式解读城市发展中的环境问题
Losing Year (YANG Teng) addresses environmental issues in the process of urban development in a childlike way

《城市森林》（陆健）采用城市中拆掉的旧建筑的木结构，开启并诉说着对城市发展的种种思考和情怀
Urban Forest (LU Jian) uses wooden materials recycled from demolished buildings to recount the story of urban development

《顺，逆》（齐婧雯）展现了经历过爆破、敲打、切割、堆砌的石材，从大山挪移到了城市的建筑中，同时诉说着它与这座城市密不可分的羁绊
Direct, Inverse (QI Jingwen) shows stones that have been blasted, tapped, cut, and stacked, then moved from mountains to buildings, and tells the close relationship between the material and the city

上图+下图：徐家汇连廊希望通过连接容纳各功能的建筑单体，布局多层次的公共空间体系，增加区域公共开放空间

Xujiahui Skywalk project aims to connect buildings with various functions in order to develop a multi-level public space system and increase public spaces in the district

「城市微介入·音乐奏响社区」：虹口区音乐谷实践案例展

URBAN MICRO-INTERVENTION, MUSICIFY THE COMMUNITY: SITE PROJECT OF SHANGHAI MUSIC VALLEY, HONGKOU

EXHIBITION THEME: URBAN MICRO-INTERVENTION, MUSICIFY THE COMMUNITY

策展主题：「城市微介入·音乐奏响社区」

项目简介

本届虹口区实践案例展接续2015上海城市空间艺术季虹口案例，在2017年延续2015年的“城市微空间复兴计划”（Let's work）项目，将音乐谷社区、提篮桥街道作为两大展区，邀请同济大学设计创意学院、知名建筑事务所等机构院校，在音乐谷地区重点打造音乐创客产业，开展三角地园区微改造；打造音乐谷社区博物馆；为音乐谷设计完善的步行服务系统，重点提升音乐谷地区的社区品质；策划一系列社区观察项目，收集音乐谷社区在转型与更新进程中的一手资料，搭建融入场地的社区博物馆；激活提篮桥历史文化风貌区内的犹太教堂、住宅、店铺、学校等建筑物及犹太人的生活印记，以提升虹口区甚至整个上海独特的历史文化影响力。

策展主题阐释

“城市微空间复兴计划”旨在发动具有社会使命感的设计师们“寻找和研究城市失落空间并进行设计”，激活负面空间，让生存空间更积极，让社区空间更有活力；也让这些“失落空间”能够转变成市民个人生活与公共活动的重要载体，唤起市民对所居住社区的归属感。这项计划目前已得到广大中青年建筑师的积极响应。本次虹口案例展延续2015年提出的“小步伐，大改变”的整体理念，提出“微介入、大提升”的实践思路，在公共空间策划中引入先锋设计与艺术思维，使实践作品从之前的单点个案拓展为网络化、互动化、多元参与的空间艺术实践。

PROJECT INTRODUCTION

Continuing the project of "Urban Humble-Space Regeneration Plan" (Let's work) in 2015, this year's Hongkou Site Project is an extension of 2015 SUSAS. The exhibition selects Music Valley Community and Tilanqiao neighborhood as two major exhibition areas. Institutions like College of Design and Innovation of Tongji University and architectural studios were invited to create a music-maker industry in Music Valley, carrying out micro-innovation at Sanjiaodi Art Park. Other micro interventions include creating a community museum; designing a complete walking service system to improve the quality of community; carrying out a series of community observation projects to collect first-hand information of Music Valley during its process of transformation and renewal with building a museum that integrates into the site; rehabilitating the old buildings in Tilanqiao historical and cultural area, such as synagogues, homes, shops, schools, as well as the Jewish life imprint, to enhance the unique historical and cultural influence of Hongkou and broadly in Shanghai.

THEME INTERPRETATION

"Urban Humble-Space Regeneration Plan" intends to call for designers with a sense of social responsibility to "find and study the lost spaces in the city and further conduct new design". It aims to activate passive spaces, and to make the living space more lively and the community more vibrant. Hopefully these "lost spaces" can be transformed to an important carrier of citizens' individual life and public activities, increasingly arousing the public's sense of belonging to their community. This plan has been welcomed by junior and intermediate architects. Based on the overall concept of "Small Steps, Big Change" proposed in 2015, this year's site project brought up the practice guideline of "Micro Intervention, Massive Improvement", following which the pioneering designs and artistic thinking were introduced to public space planning. It expands practical works from the previous context of singular cases to networked and interactive multi-participation.

展览基本信息

主题：城市微介入·音乐奏响社区

时间：2017.11—2018.1

举办地址：

社区博物馆、社区体验馆：嘉兴路260号三角地艺术园二层

音乐市集、水岸景观：音乐谷片区辽宁路段+沙泾港

街区微改造：提篮桥街区

主办单位：虹口区人民政府

承办单位：虹口区规划和土地管理局、Let's Talk论坛

协办单位：上海音乐谷集团有限公司、提篮桥街道办公室、虹口文化创意产业推进领导小组办公室

策展团队介绍

戴春

博士，《时代建筑》杂志责任编辑／运营总监，Let's Talk论坛创办人，城市微空间复兴计划发起人，筑术空间主持人。

俞挺

Wutopia Lab主持建筑师，东南大学客座教授，城市微空间复兴计划发起人，Let's Talk论坛创办人。

Let's talk学术论坛

开创于2014年底的Let's talk学术论坛由俞挺和戴春共同创立，三年来，已成功举办近120场专题演讲与讨论，是沪上充满活力的公共热点。论坛的一系列讨论深刻触及城市与建筑问题。作为论坛的创始人和组织者，他们在2015年又发起"城市微空间复兴计划"，号召设计师们为身边的城市空间做些什么，让Let's talk变身为Let's work，透过充满智慧且在实施上简便易行的设计构思，让那些在公共生活中的"失落空间"能够迅速转变成市民个人生活与公共活动的重要载体，唤起每个市民对所居住社区更多的归属感。

Basic Information

Theme: Urban Micro Intervention, Musicify the Community

Time: November 2017 – January 2018

Venue: Community Museum/Experience Hall: Second Floor, Sanjiaodi Art Park, Music Valley (No. 260 Jiaxing Road)

Music Valley/Music Fair: Waterside sceneries (Liaoning Road and Shajinggang Creek sections of Music Valley)

Micro-renovation: Tilanqiao Community

Sponsor: Hongkou District Government

Undertaker: Hongkou District Urban Planning & Land Authority, Let's Talk forum

Supporters: Shanghai Music Valley Co., Ltd,
Sub-district Office of Tilanqiao,
Hongkou District Lead Team Office for Promotion of Cultural Creativity Industry

Curators

DAI Chun

Ph.D., Editor & Operation Director of *Time Architecture* magazine; Co-founder of Let's Talk Forum; Initiator of "Urban Humble-Space Regeneration Plan"; Principal of Archiepos Studio.

YU Ting

Chief architect of Wutopia Lab; Guest professor at Southeast University; Launcher of " Urban Humble-Space Regeneration Plan"; Co-founder of Let's Talk forum.

Let's Talk Academic Forum

As the most active inoffical acdemic forum in Shanghai, Let's talk focuses on architecture, art and culture communication. Since founded in 2014, nearly 120 events were held. The forum has already become the most influencial one of Shanghai.

In 2015, Yu Ting & Dai Chun, co-founders of Let's Talk academic forum, launched the "Urban Humble-Space Regeneration Plan", calling the designer to examine their neighboring space and carry out micro-regeneration practices. The plan aims to activate negative space to form positive living space and dynamic community space; thus, the "humble space" can be transformed into important carriers of private life and public activities for citizens. More importantly, it is to arouse every citizen's sense of belonging to his or her living community.

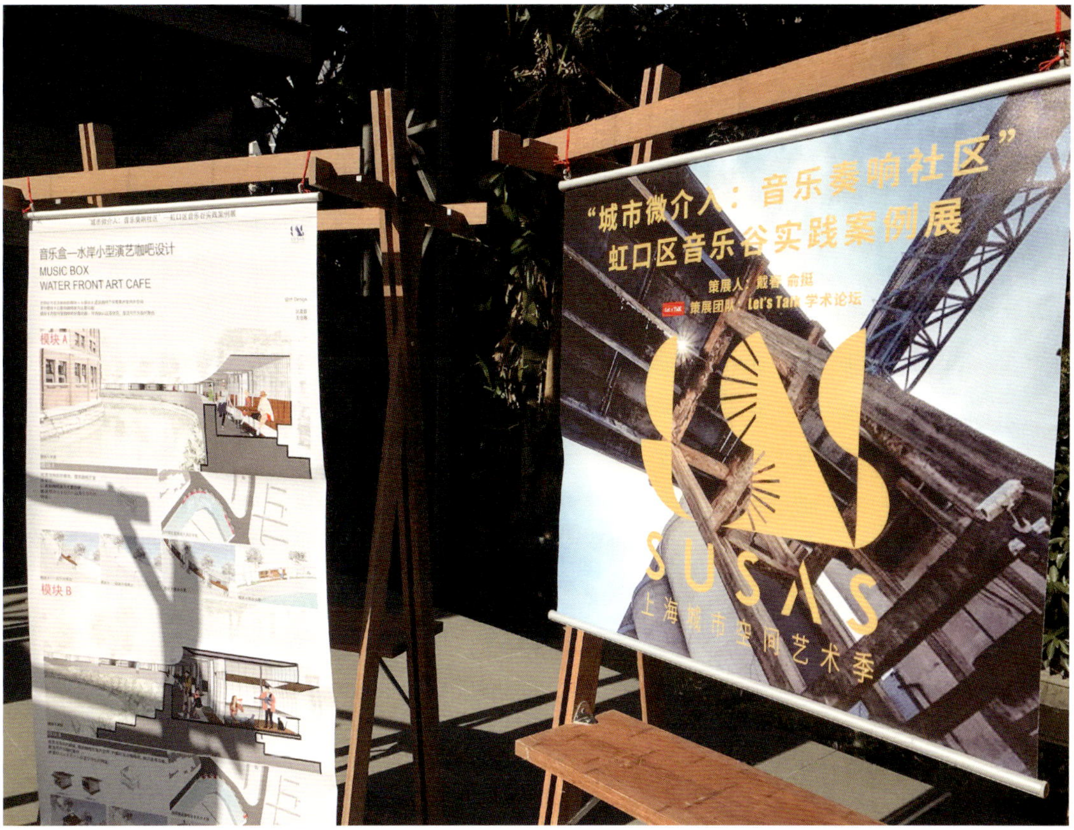

向公众展示改造设计的音乐谷展览现场
Music Valley exhibition site

实践案例索引

序号	案例名称	策展人及团队	案例地址	简介
1	社区，让生活更美好：浦东新区缤纷社区实践案例展	奚文沁、徐继荣、杨帆	浦东新区陆家嘴街道、洋泾街道、潍坊街道、塘桥街道、花木街道	浦东新区缤纷社区实践案例展包括了1部宣传片、1个空间艺术展览、1+5“缤纷沙龙”、1个低碳主题活动和1次摄影比赛，邀请5个街道与多方部门共同参与，呈现缤纷社区的规划成果。
2	新泾·新境：长宁区北新泾街道实践案例展	张佳晶、周渐佳	长宁区北渔路95号民俗文化中心南楼	长宁区北新泾街道实践案例展主要由“更新进入新泾”“展览进入社区”“学术进入日常”三个板块组成，集中展示微更新与城市系统更新两方面的举措与设想。
3	“符号上海”·上海风貌保护道路文化之旅：上海城市规划展示馆实践案例展	翁文斌	上海城市规划展示馆（人民大道100号）	“符号上海”作为上海城市规划展示馆聚焦城市历史风貌遗产的品牌展示活动项目，其创新了将规划专业展览与艺术创作和公众参与活动相结合的“众筹”的展览模式。
4	为风貌而设计：徐汇区岳阳路实践案例展	薛鸣华、王林	徐汇区复兴中路1331号黑石公寓、永嘉路578号、永嘉路511号	岳阳路实践案例展由风貌影像展、历史文化展、天平名人展、互动设计音乐展等内容组成，通过公共空间的艺术语言，建立人与物之间更直接的关系。
5	双重曝光·旧厂新生：静安区新业坊实践案例展	上海罗浮紫艺术品有限公司	静安区汶水路210号静安新业坊文化创意产业基地	静安新业坊实践案例展以“咱们工人有力量”和“咱们人人有力量”为主题，配合艺术装置、艺术市集和音乐派对等艺术活动，使市民切身感受城市空间通过在地化创作和公共艺术实践带来的提升及改变。
6	“双向辐射”·城市更新在西岸：徐汇区西岸实践案例展	陈安达、叶可央、杨丹	徐汇西岸公共开放空间（跑道公园、星美术馆、油罐艺术中心、西岸艺术中心等周边区域）	徐汇区西岸实践案例展主要内容为联动西岸艺术季内相关展览和公众活动，展示内容包括：星美术馆、跑道公园、滨江公共开放空间内的十余个公共艺术作品。
7	融“汇”贯通：徐汇区T20大厦实践案例展	傅春禹、上海天华城市规划设计有限公司	徐汇区天钥桥路20号T20大厦（原西亚宾馆）	T20大厦实践案例展从城市建设发展、艺术文化发展的实际出发，结合作品展览与系列文化活动等多种形式，展出徐家汇城市空间若干实践案例和不同领域艺术家的作品。
8	“城市微介入·音乐奏响社区”：虹口区音乐谷实践案例展	戴春、俞挺	虹口区音乐谷片区、提篮桥街道	虹口案例展活动在公共空间策划中引入先锋设计与艺术思维，使实践作品从之前的单点个案拓展到网络化、互动化、多元参与的空间艺术实践。

参与单位	案例形式	呈现内容
上海市浦东新区人民政府、上海市浦东新区规划和土地管理局、上海市城市规划设计研究院、上海市浦东新区规划设计研究院、中国（上海）自由贸易试验区管理委员会陆家嘴管理局、浦东新区陆家嘴街道、潍坊街道、洋泾街道、塘桥街道、花木街道	4项活动	缤纷社区行动计划成果发布暨社区规划主题沙龙、“低碳出行”主题活动、街道系列沙龙、“美丽家园”摄影比赛
长宁区人民政府、长宁区规划和土地管理局、长宁区北新泾街道、长宁区建设和交通委员会、长宁区文化局	6项活动	开幕式、“天山西路-城市慢行”专题论坛、青少年社区行走、“新泾诸村-社区新生”专题论坛、露天电影、北新泾字体工作坊
上海城市规划展示馆	10个展览活动	符号上海——街巷·童趣、符号上海——上海风貌保护道路文化之旅、悦读城市——永不拓宽的街道、符号上海——街巷·彩韵、街巷·活力——城市定向赛、留住城市的记忆——航拍历史文化风貌区摄影展、悦读城市——规划科普巡展、画境之外——城市与乡间的时空印迹、连接——城市公共空间的多元视角　同济大学建筑与城市规划学院国际教学实验展、符号上海——街巷·经纬
徐汇区人民政府、徐汇区规划和土地管理局、上海交通大学城市更新保护创新研究中心、上海安墨吉建筑规划设计有限公司、上海衡复投资发展有限公司	展览+活动	黑石影像展、“来自法国的上海人”赉安洋行建筑作品在徐汇、“海派京剧名家”天平名家坊、“声·相”城市风貌音乐周
静安区人民政府、静安区规划和土地管理局、静安新业坊文化创意产业基地	2场公众活动	“咱们工人有力量”艺术活动和“咱们人人有力量”艺术活动
徐汇区人民政府、徐汇区规划和土地管理局、上海西岸集团、上海光启文化产业投资发展有限公司	展览	星美术馆、跑道公园、油罐艺术中心、滨江公共开放空间内的十余个公共艺术作品及相关展墙
徐汇区人民政府、徐汇区规划和土地管理局、T20大厦（原西亚宾馆）	以展览为主	艺术家的作品展览，包括张海平、杨冬白、蒋飞、王静艳、尹代波、李乾煜、陆健、田君宇、朱屹立、朱晓雯、漆佩松、杨腾、张灵、齐婧雯、翁泽剑、吴昊等
虹口区人民政府、虹口区规划和土地管理局、Let's Talk学术论坛、上海音乐谷集团有限公司、提篮桥街道办公室、虹口文化创意产业推进领导小组办公室	展览	公共空间展示

SITE PROJECTS INDEX

No.	Exhibition Title	Curators	Venue	Introduction
1	Better Community, Better Life: Site Project of Pudong Colorful Community	XI Wenqin, XU Jirong, YANG Fan	Lujiazui, Yangjing, Weifang, Tangqiao and Huamu Communities in Pudong New Area	This exhibition includes one promotion film, one space art exhibition, "Colorful Salon" (one main forum+ five sub-forums), one low-carbon theme event and one photography competition. Five sub-district administrations and other authorities cooperated, showcasing the planning outcomes of Colorful Community.
2	Xinjing, New Scape: Site Project of Changning Beixinjing Community	ZHANG Jiajing, ZHOU Jianjia	1st Floor & 2nd Floor, South Building, Changning Folk Culture Center (No. 95 Beiyu Road, Changning District)	The site project of regeneration in Beixinjing, Changning is composed of three parts, namely "Urban Regeneration in Xinjing", "Exhibition in Community" and "Academy in Everyday Life", which highlights conceptions and implementations at two scales: communal amenities and urban system.
3	2017 "Iconic Shanghai": Serial Exhibitions at Shanghai Urban Planning Exhibition Center	WENG Wenbin	Shanghai Urban Planning Exhibition Center (No.100 People Avenue, Huangpu District)	Hosted by Shanghai Urban Planning Exhibition Center, "Iconic Shanghai" focuses on the historical landscapes of the city. The serial events creatively adopt the "crowd funding" model that combines curated exhibition, art work production and public participation.
4	Design for the Historic and Cultural Scene: Site Project of Yueyang Road, Xuhui	XUE Minghua, WANG Lin	Blackstone Apartments (No. 1331 Middle Fuxing Road); No. 578 and 511 Yongjia Road, Yongjia New Village	The site project emphasizes the timeline and humanistic evolution of public space, showcasing features and transformation of contemporary humanities, life, and landscape in a holistic manner.
5	Double Exposure, Factory Rebirth: Site Project of Innovation Galaxy, Jing'an	Shanghai Purple Roof Art Gallery	Jing'an Innovation Galaxy (No.210 Wenshui Road, Jing'an District, Shanghai)	Jing'an Innovation Galaxy launches two sessions of Art Exhibitions, "We Workers Have Power" and "We All Have Power ", together with participatory and interactive art installations, art markets and music parties. These localized public events are designed to inspire visitors with the concept that art transforms and sublimates our beloved city.
6	Dual Radiation: Site Project of Urban Renewal at West Bund, Xuhui	CHEN Anda, YE Keyang, YANG Dan	West Bund Public Open Space (Start Museum, TANK Shanghai, West Bund Art Center, and surrounding)	This project aims to raise the public consciousness of issues like management of waterfront public spaces, art in public spaces and preservation of industrial heritages.
7	Linking Zikawei: Site Project of T20 Building, Xuhui	FU Chunyu, Shanghai Tianhua Urban Planning & Design Co., Ltd	No. 20 Tianyaoqiao Road	Considering the existing conditions of urban construction and cultural development, this project adopts exhibitions and serial events to showcase a number of site cases in Xujiahui urban space, and art works composed by a variety of artists.
8	Urban Micro-Intervention, Musicify the Community : Site Project of Shanghai Music Valley, Hongkou	DAI Chun, YU Ting	Music Valley and Tilanqiao Community in Hongkou District	The project introduces experimental design and art thinking into public space planning while extending isolated sites into an interactive, participatory grid of space art.

Organizers	Forms	Events
People's Government of Pudong New Area, Shanghai; Pudong New Area Planning and Land Authority; Shanghai Urban Planning and Design Research Institute; Pudong New Area Planning and Design Research Institute; Lujiazui Office of China (Shanghai) Pilot Free Trade Zone Administration; Sub-district offices of Lujiazui, Weifang, Yangjing, Tangqiao, Huamu Communities	Four activities	Theme Salon: Achievements and Planning of Colorful Community; Theme Event: Low-carbon Travelling; Serial Salons: Our Streets; Photography Competition: Our Beautiful Home
People's Government of Changning District, Shanghai; Changning District Planning and Land Authority; Sub-district Office of Beixinjing, Changning District; Construction and Traffic Committee of Changning District; Culture Bureau of Changning District	Six activities	Opening Ceremony; Theme Forum: West Tianshan Road – Slow Walk in City; Community Walk Tour for Teenagers; Theme Forum: Xinjing Villages – Community Revitalization; Outdoor Movies; Typeface Workshop in Beixinjing
Shanghai Urban Planning Exhibition Center	Ten exhibitions and activities	Streets of Childish Joy: Painting Competition for Children; Streets and Heritage: Workshops and Lectures on Shanghainese Craftsmanship; Streets in Perception: "Pleasant Reading of City" series lectures; Streets and Color; Streets Energized: Urban Orienteering; Streets behind Flashlight: Photography Competition for Everyone; Pleasant Reading of City–Tour Exhibition of City Planning; Beyond the Painting – the time and space imprint of cities and the countryside; Teaching Experiment Exhibition in Tongji University School of Architecture and Urban Planning about Multi-perspective view on urban public space; Latitude and Longitude of Streets
People's Government of Xuhui District, Shanghai; Xuhui District Planning and Land Authority; Shanghai Jiao Tong University City Lab; Shanghai AMJ Architecture & Urban Planning Design Co. Ltd; Shanghai Hengfu Investment & Development Co., Ltd	Exhibitions and public activities	Landscape Photography Exhibition, History & Culture Exhibition, Celebrities of Tianping Street, Interactive Design Music Show
People's Government of Jing'an District, Shanghai; Jing'an District Planning and Land Authority; Jing'an Innovation Galaxy	Two public art exhibitions	Two sessions of Art Exhibitions: "We Workers Have Power" and "We All Have Power "
People's Government of Xuhui District, Shanghai; Xuhui District Planning and Land Authority; Shanghai West Bund Development (Group) Co., Ltd; Shanghai Guangqi Cultural Industry Investment Development Co., Ltd	Exhibitions	The most important exhibited items include: Start Museum, Runway Park, TANK Shanghai and more than ten pieces of public art in the Public Open Space.
People's Government of Xuhui District, Shanghai; Xuhui District Planning and Land Authority; T20 Building (formerly West Asia Hotel)	Exhibitions	Art works from ZHANG Haiping, YANG Dongbai, JIANG Fei, WANG Jingyan, YIN Daibo, LI Qianyu, LU Jian, TIAN Junyu, ZHU Yili, ZHU Xiaowen, QI Peisong, YANG Teng, ZHANG Ling, QI Jingwen, WENG Zejian, WU Hao
People's Government of Hongkou District, Shanghai; Hongkou District Planning and Land Authority; Let's Talk; Shanghai Music Valley Group Co., Ltd; Sub-district office of Tilanqiao; Hongkou District Lead Team Office for Promotion of Cultural Creativity Industry	Exhibitions	Public space exhibitions

SUSAS
Shanghai Urban Space Art Season 2017
2017 上海城市空间艺术季
连接 thisCONNECTION
共享未来的公共空间
Sharing a Future Public Space
时间 HOLDING TIME
2017 年 10 月 15 日至 2018 年 1 月 15 日
Oct 15, 2017 to Jan 15, 2018
地点 MAIN EXHIBITION VENUE
主展场民生码头 8 万吨筒仓及周边开放空间
主办 上海市城市雕塑委员会
承办 上海市规划和国土资源管理局
上海市浦东新区人民政府

SUSAS
上海城市空间艺术季
Shanghai Urban Space Art Season 2017
2017上海城市空间艺术季
缤纷内城 漫步浦东
社区，让生活更美好——浦东缤纷社区实践案例展
Better Community, Better Life—Site Project of Pudong Colorful Community
展览时间
Exhibition Time
2017年10月15日——2018年1月15日
October 15, 2017 to January 15, 2018
展览地点
Exhibition Venue
民生码头8万吨筒仓7楼（民生路3号）
7F, No.3 Minsheng Road

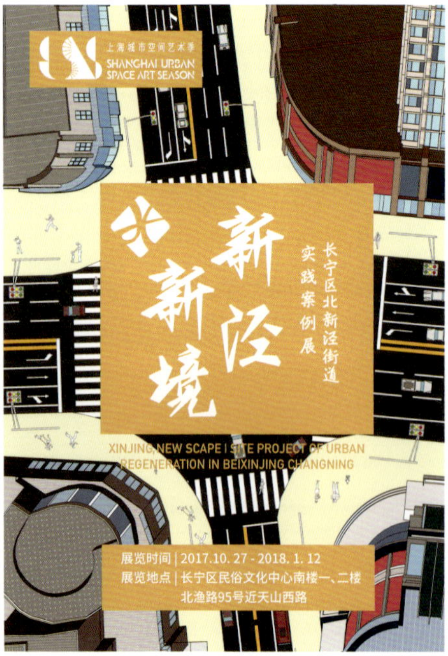
上海城市空间艺术季
SHANGHAI URBAN SPACE ART SEASON
新新境
新泾
长宁区北新泾街道
实践案例展
XINJING NEW SCAPE I SITE PROJECT OF URBAN REGENERATION IN BEIXINJING CHANGNING
展览时间 | 2017. 10. 27 - 2018. 1. 12
展览地点 | 长宁区民俗文化中心南楼一、二楼
北渔路95号近天山西路

SUSAS
符号上海
ICONIC SHANGHAI
SHANGHAI FEATURED ROAD
上海风貌保护道路
文化之旅
CULTURAL TOUR
系列展览
SERIES EXHIBITION

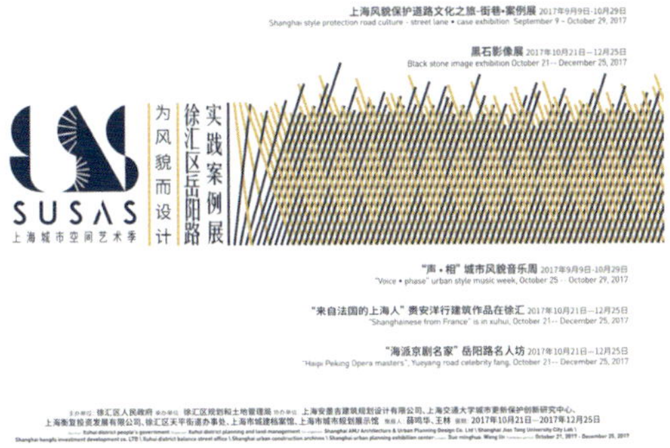
上海风貌保护道路文化之旅-街巷•案例展 2017年9月9日-10月29日
Shanghai style protection road culture - street lane • case exhibition September 9 - October 29, 2017
黑石影像展 2017年10月21日-12月25日
Black stone image exhibition October 21 - December 25, 2017
SUSAS
上海城市空间艺术季
徐汇区岳阳路实践案例展
为风貌而设计
"声•相"城市风貌音乐周 2017年9月9日-10月29日
"Voice • phase" urban style music week, October 25 - October 29, 2017
"来自法国的上海人"贵安洋行建筑作品在徐汇 2017年10月21日-12月25日
"Shanghainese from France" is in xuhui, October 21 - December 25, 2017
"海派京剧名家"岳阳路名人坊 2017年10月21日-12月25日
"Haipi Peking Opera masters", Yueyang road celebrity fang, October 21 - December 25, 2017

SUSAS
上海城市空间艺术季
Shanghai Urban Space Art Season 2017
2017上海城市空间艺术季
静安新业坊实践案例展
Jing'an Innovation Galaxy Site Project
双重曝光
旧厂新生
Double Exposure · Factory Rebirth
时间 TIME 2017.10.21/2018.1.6
地点 VENUE 上海市静安区新业坊
（静安区汶水路210号新业坊）

Shanghai Urban Space Art Season 2017
SUSAS
上海城市空间艺术季
Shanghai Urban Space Art Season 2017
2017 上海城市空间艺术季
双向辐射 Dual Radiation
城市更新在西岸，徐汇区实践案例展
Site project of Urban Renewal at West Bund, Xuhui
时间 TIME
2017年11月至2018年1月
Nov. 2017 to Jan. 2018
地点 MAIN EXHIBITION VENUE
上海西岸
Shanghai West Bund

上海城市空间艺术季
SHANGHAI URBAN SPACE ART SEASON
融"汇"贯通 LINKING ZIKAWEI
徐家汇T20大厦实践案例展
SITE PROJECT OF T20 BUILDING, XUJIAHUI
时间 2017年11月至12月底
地点 徐汇区徐家汇T20大厦二层空间
天钥桥路20号
T20
BUILDING

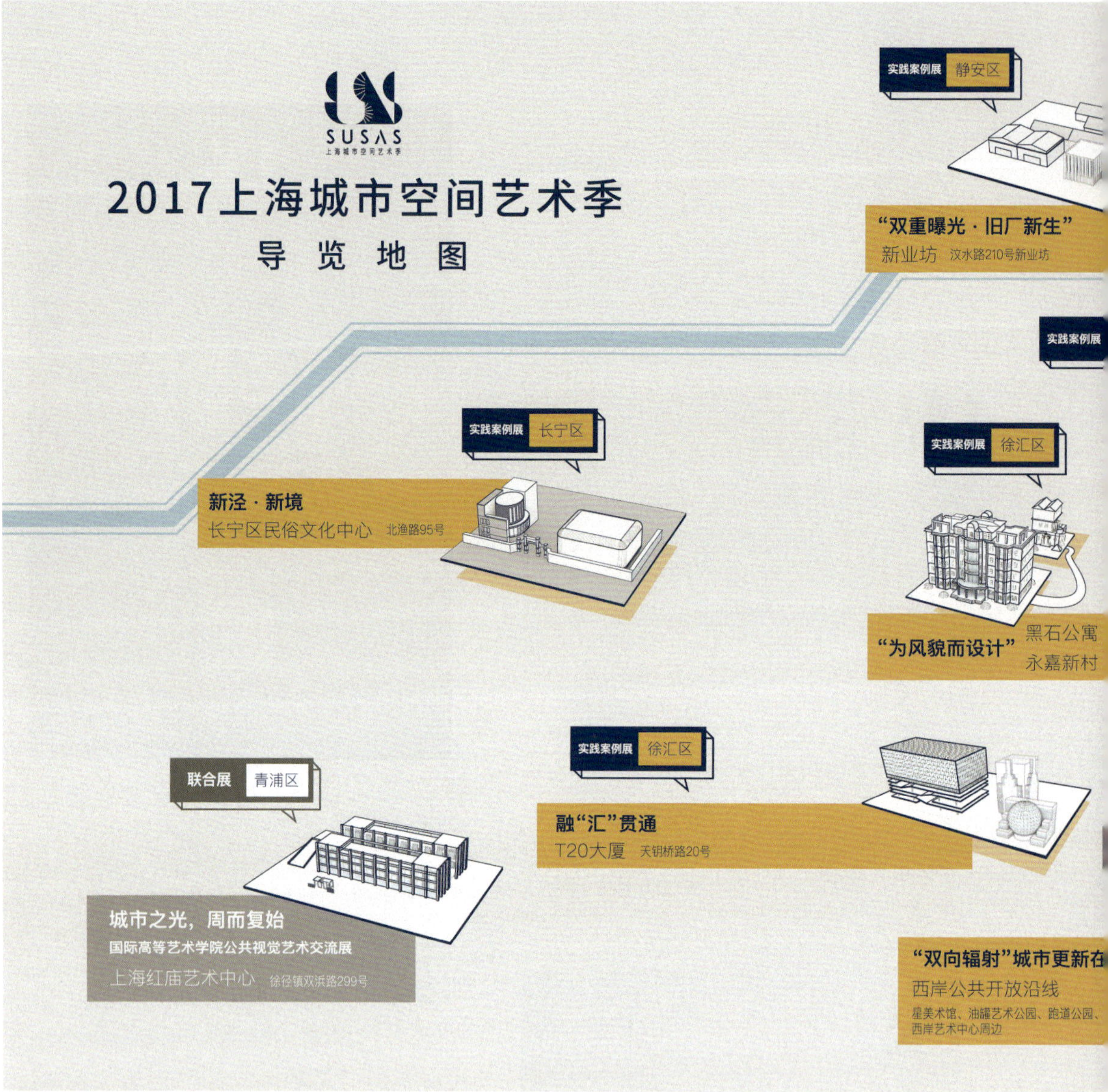

上海城市空间艺术季执行团队

总执行团队

上海市规划和国土资源管理局风貌处|侯斌超、林磊、应孔晋

上海市文化广播影视管理局艺术处|沈捷、沈雪江

上海城市公共空间设计促进中心 |赵宝静、王明颖、陈敏、胡颖蓓、王夏娴、周敏、马宏、陈成、薛娱沁、赵起超、严帅帅、章舒雯

主展览执行团队

浦东新区规划和土地管理局|关也彤、刘伟、汤明华、魏文、赵波

上海东岸投资（集团）|黄磊、程世红、韩莹赟、王芳、屈伯禹、曾雨佳、陈亮、孙政

案例展执行团队

浦东新区规划和土地管理局|关也彤、刘伟、汤明华、魏文、赵波

长宁区规划和土地管理局|郭海、苏立琼

长宁区北新泾街道|顾耀军、张生义、宋妮妮、刘薇

上海城市规划展示馆|翁文斌、曹夏雯

徐汇区规划和土地管理局|高世昀、范润生、戴炯玥

静安区规划和土地管理局|董瑜、吴炳怀、桂寅

上海西岸集团|陈安达、叶可央、杨丹

虹口区规划和土地管理局|罗翌弘、赵思齐

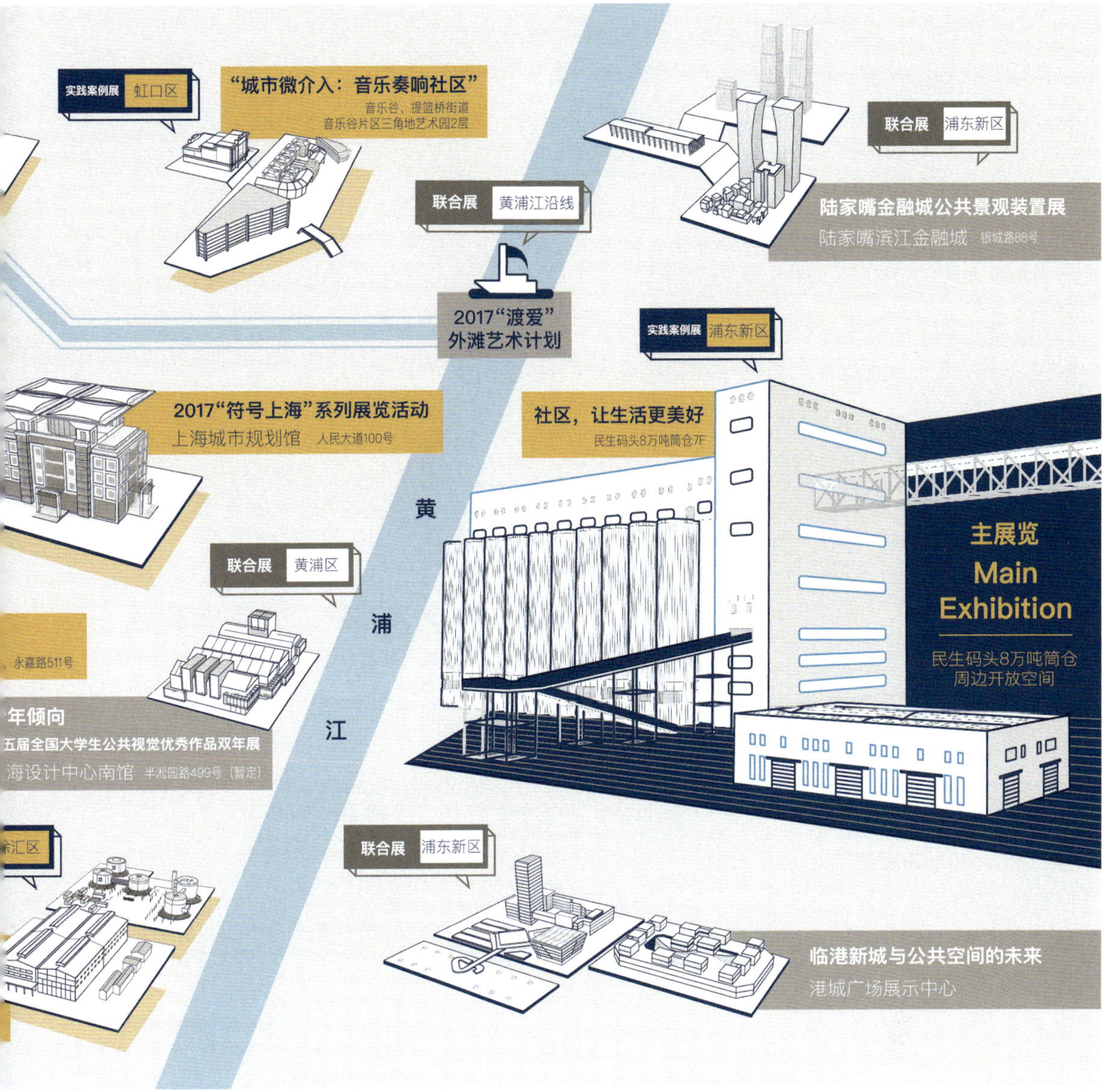

SUSAS Execution Team

GENERAL EXECUTION TEAM

Landscape Division, Shanghai Municipal Bureau of Planning and Land Resources | HOU Binchao, LIN Lei, YING Kongjin

Art Division, Shanghai Municipal Bureau of Culture, Radio, Film & TV| SHEN Jie, SHEN Xuejiang

Shanghai Design & Promotion Center for Urban Public Space| ZHAO Baojing, WANG Mingying, CHEN Min, HU Yingbei, WANG Xiaxian, ZHOU Min，MA Hong, CHEN Cheng, XUE Yuqin, ZHAO Qichao, YAN Shuaishuai, ZHANG Shuwen

MAIN EXHIBITION EXECUTION TEAM

Pudong New Area Planning and Land Authority

Shanghai| GUAN Yetong, LIU Wei, TANG Minghua, WEI Wen, ZHAO Bo

Shanghai East Bund Investment (Group)CO. ,LTD| HUANG Lei, CHENG Shihong, HAN yingyun , WANG fang, QU Boyu, ZENG Yujia, CHEN Liang，SUN Zheng

SITE PROJECT EXECUTION TEAM

Pudong New Area Planning and Land Authority

Shanghai|GUAN Yetong, LIU Wei, TANG Minghua, WEI Wen, ZHAO Bo

Changning District Planning and Land Authority, Shanghai | GUO Hai, SU Liqiong

Beixinjing Sub-district|GU Yaojun, ZHANG Shengyi, SONG Nini, LIU Wei

Shanghai Urban Planning Exhibition Center| WENG Wenbin, CAO Xiawen

Xuhui District Planning and Land Authority, Shanghai | GAO Shiyun, FAN Runsheng, DAI Jiongyue

Jing'an District Planning and Land Authority, Shanghai | DONG Yu, WU Binghuai, GUI Yin

Shanghai West Bund Development(Group)Co.,Ltd | CHEN Anda, YE Keyang, YANG Dan

Hongkou District Planning and Land Authority, Shanghai| LUO Yihong, ZHAO Siqi

图书在版编目（CIP）数据

连接：共享未来的公共空间实践：2017上海城市空间艺术季案例展 / 上海城市空间艺术季展览画册编委会编. -- 上海：同济大学出版社，2019.8
ISBN 978-7-5608-8723-4

Ⅰ. ①连… Ⅱ. ①上… Ⅲ. ①城市规划 – 空间规划 – 上海 – 画册 Ⅳ. ①TU984.251-64

中国版本图书馆CIP数据核字(2019)第185622号

连接：共享未来的公共空间实践
thisCONNECTION: SHARING A FUTURE PUBLIC SPACE SITE PROJECT

2017上海城市空间艺术季案例展
2017 SHANGHAI URBAN SPACE ART SEASON SITE PROJECT

上海城市空间艺术季展览画册编委会 编
EDITED BY: SHANGHAI URBAN SPACE ART SEASON ALBUM EDITORIAL COMMITTEE

出 版 人：华春荣
策　　划：秦蕾 / 群岛工作室
项目统筹：马宏　王夏娴
责任编辑：李争
责任校对：徐春莲
装帧设计：付超
版　　次：2019年8月第1版
印　　次：2019年8月第1次印刷
印　　刷：上海雅昌艺术印刷有限公司
开　　本：787mm × 1092mm 1/16
印　　张：12
字　　数：299 000
书　　号：ISBN 978-7-5608-8723-4
定　　价：128.00元
出版发行：同济大学出版社
地　　址：上海市四平路1239号
邮政编码：200092
网　　址：http://www.tongjipress.com.cn
本书若有印装问题，请向本社发行部调换

ISBN 978-7-5608-8723-4
Publisher: HUA Chunrong
Initiated by: QIN Lei / Studio Archipelago
Project coordinator: MA Hong / WANG Xiaxian
Editor: LI Zheng
Proofreading: XU Chunlian
Graphic Design: FU Chao
Published in August 2019, by Tongji University Press, 1239, Siping Road, Shanghai, China, 200092.
www.tongjipress.com.cn

上海城市空间艺术季 http://www.susas.com.cn
光明城联系方式：info@luminocity.cn

SUSAS http://www.susas.com.cn
Contact us: info@luminocity.cn